AIRCRAFT PROPULSION

AIRCRAFT PROPULSION

MAYUR R ANVEKAR
Assistant Professor
Department of Aeronautical Engineering
Nitte Meenakshi Institute of Technology (NMIT)
Bengaluru

PHI Learning Private Limited
Delhi-110092
2016

₹ 175.00

AIRCRAFT PROPULSION
Mayur R Anvekar

ISBN-978-81-203-5264-3

Published by Asoke K. Ghosh, PHI Learning Private Limited, Rimjhim House, 111, Patparganj Industrial Estate, Delhi-110092 and Printed by Raj Press, New Delhi-110012.

To

My Parents, Rohini and Rajendra

and

Sister, Manali

for their patience and love

CONTENTS

PREFACE

I am happy to come out with this book *Aircraft Propulsion*. The material in this book mainly supports the undergraduate course on Aircraft Propulsion. I spent a considerable amount of time in writing this book, which contains all the major topics of propulsion course required for an undergraduate student and any beginner.

Chapter 1 in the book explains about the fundamental concepts required for propulsion and various propulsion systems. Gas turbine engine components are theoretically discussed in detail in Chapters 2, 3 and 4. Rocket propulsion, ramjet engine and scramjet engine are explained in Chapters 5 and 6. The pulse detonation engine is a research-oriented topic, which is introduced at the basic level in Chapter 7.

My intention was not to write a book that was only exam-oriented, but one that will help the readers to improve their understanding of the subject. Hopefully, this will make it easier for the students and other beginners to understand the basics of propulsion. I have suggested a list of books at the end, which will give an in-depth understanding of the subject.

I am grateful to Dr. Vivek Sanghi for contributing his time and effort in reviewing the book. I have benefited from his valuable comments and suggestions.

I wish to express my gratitude to my parents for all the support. I especially thank my sister, Manali for all the inspiration and motivation.

Mayur R Anvekar

CHAPTER 1

INTRODUCTION

OVERVIEW

This chapter begins with the basic concepts and laws of thermodynamics, heat transfer, governing equation and Newton's laws of motion. In this, chapter we will look into the thermodynamic cycle used for different types of propulsion systems and their thrust force generation. Almost all the basic types of propulsion systems are explained to the basic level of understanding. We will also look into some sections like accessory section, auxiliary power unit, engine operation and noise in this chapter.

Aircraft propulsion or *aero propulsion* is a science which deals with aero engines and its sub-components. It is also known as the branch of aeronautics and aerospace, which deals with the aircraft and spacecraft powerhouse system.

Aircraft and spacecraft are the flying vehicles which are heavier than air, and propulsion means moving forward or study of propelling. To make the aircraft in motion, we need a force, which is generated by the momentum of gases or air with some pressure and velocity. This momentum of gases or air is produced using the propulsive devices or aero engines. The variation in kinetic energy and pressure energy of the fluid across the engine components leads to the required force generation. For example, when a balloon is blown, the balloon material gets stretched and starts resisting on more expansion, which leads to increase in pressure of air inside the balloon. The moment we free the balloon, the air which is inside the balloon comes out as a stream at high velocity, which leads to the motion of balloon in opposite direction to the air stream. This action and reaction lead to propelling of the balloon.

In 250 BC, a mathematician named Hero devised a toy called *aeolipile*, which consists of a modified kettle, with two vertical tubes on the periphery of the kettle. The kettle is round in shape and supported from two ends perpendicular to the vertical tubes such that it can be easily rotated, as shown in Figure 1.1. The kettle is filled with water and continuously heated up with an external heat source, which makes the water present inside the kettle to evaporate. The steam of water vapours is directed through the convergent-shaped tubes in such a way that it ejects out at high velocity and gives momentum to the kettle to rotate. This is the first instrument, where the heat energy or steam energy is transformed into mechanical energy or kinetic energy.

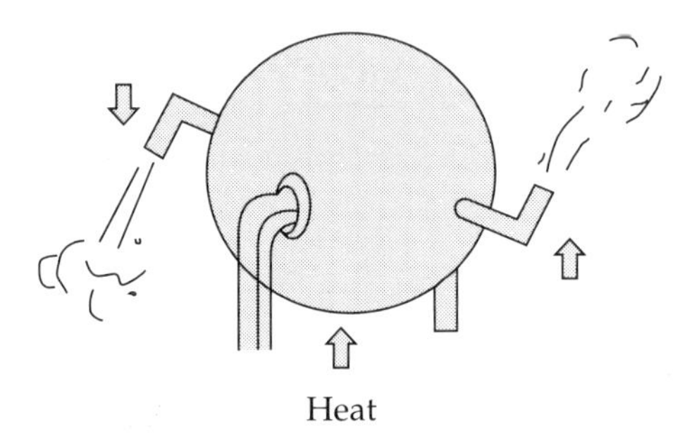

Figure 1.1 Aeolipile.

The propulsion system or engine is not a simple device which can be designed and fabricated without the basic science theories and laws. The working of an engine involves fluid mechanics and dynamics, heat interactions, mechanical motions and transmissions and other phenomena of science. Hence, to study the force generation and its flowchart, the basic knowledge of science is very essential.

1.1 REVIEW OF THERMODYNAMIC PRINCIPLES

Before we start the actual propulsion system, it is essential to know and understand the fundamentals of thermodynamics because every propulsion system deals with heat energy interaction. The aircraft propulsion is a subject where we can see the basics of heat transfer, mass transfer, thermodynamics and turbomachineries from an application perspective. Heat is a form of energy generally known as thermal energy, and mass is an external property of a system. Heat and mass transfer together is a branch of science, which deals with heat and mass interactions between the system and surrounding or between two systems.

Thermodynamics is a branch of science, which deals with the study of heat and work. Thermodynamics is classified into two parts based on its approach, namely, microscopic approach and Macroscopic approach. The Microscopic thermodynamic approach involves deeper understanding at the molecular level of a system and taking a look at the individual molecular behaviour. The internal energy of a system is the example of microscopic approach or microscopic energy and it is also called *statistical thermodynamics*. The macroscopic thermodynamic approach, is an easier approach which does not require the knowledge of a system at its molecular level. Kinetic energy and potential energy are the examples of the macroscopic approach of thermodynamics or energy and it is also called *Classical thermodynamics*.

As stated earlier, thermodynamics is a branch of science, which deals with the study of heat energy interaction between systems and its effect on the properties of the system and the work done. In simple words, thermodynamics is the study of heat in motion. It gives a detailed picture of heat interactions and its effects when a hot/cold fluid or matter is in motion. The governing laws of thermodynamics, as well as some related terms are explained here.

Zeroth law of thermodynamics

Consider three systems, namely, object 1, object 2 and object 3, as shown in Figure 1.2. If object 3 is in thermal equilibrium with object 2 and object 2 is in thermal equilibrium with object 1, then we can say that object 1 is in thermal equilibrium with object 3. This law explains about the term—'equilibrium'. Thermal equilibrium means there is no exchange of heat energy between two systems or between system and surrounding.

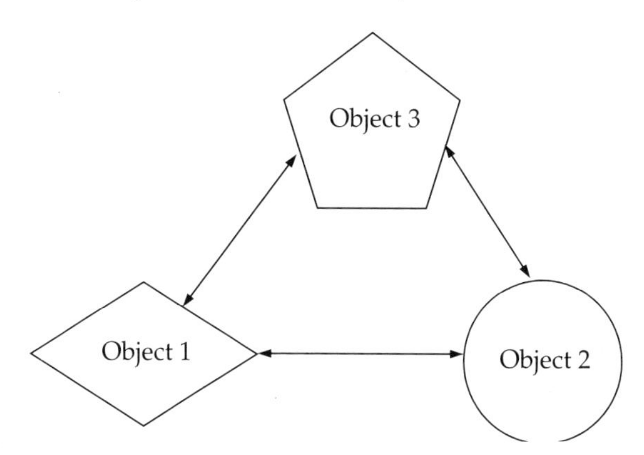

Figure 1.2 Zeroth law of thermodynamics.

First law of thermodynamics

It is also called law of conservation of energy, i.e., energy can neither be created nor destroyed, but it can be converted from one form to another, as shown in Eq. (1.1). This law explains about the energy budgeting, as shown in Figure 1.3. For example, fuel is burnt in a closed container in the presence of air, which results in an increase in pressure and temperature inside the container. This means that the chemical energy of the fuel is converted into heat energy and pressure energy of the fluid inside the container.

$$\int \partial Q \approx \int \partial W \tag{1.1}$$

where, the integral of dQ is the heat supplied and integral of dW is the work done.

$\left(\int \partial Q - \int \partial W\right)$ must be a property of system called *internal energy* (U), as shown in Eq. (1.2).

$$dU = \left(\int \partial Q - \int \partial W\right) \tag{1.2}$$

where, dU is the change in internal energy of the system.

If the heat is supplied to the system, then it will result in the same amount of work done with no loss. This is called *perpetual motion mechanism,* which is an ideal condition defined in Eq. (1.1), and (1.2) describes that for an actual condition, i.e., change between the heat supplied to the system and the work done by the system is the internal energy of the system. For example, consider a vessel with water. When heat is supplied to boil the water, initially, the heat is absorbed to raise the water temperature. That energy is the internal energy of water. Internal energy is the energy stored by virtue of its molecular motion.

The magnitude of internal energy can be changed by heating the system or by doing work or by exchanging the matter. Internal energy changes in open and closed system, but not in the isolated system.

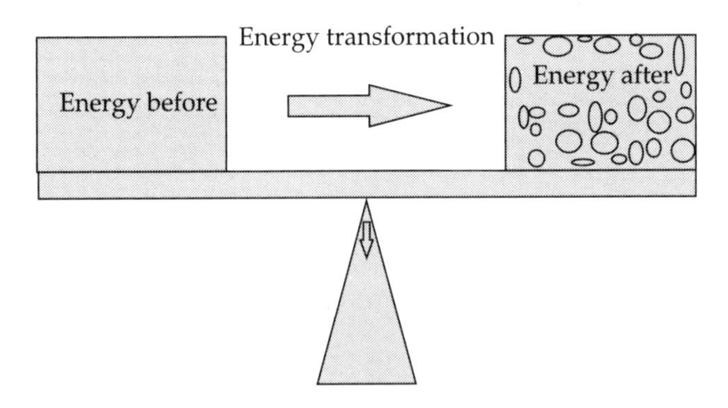

Figure 1.3 First law of thermodynamics.

Modes for storing the internal energy in the molecule are of three types, namely, translation mode, rotational mode and vibrational mode. Considering diatomic substance, *translation mode* means the molecules have linear motion in all the three dimensions and the fully excited temperature for a diatomic molecule to move in three dimensions is 5 K. *Rotational mode* is rotary motion of one atom in all the three dimensions with reference to another atom and this happens at a temperature between 5 K to 600 K. *Vibrational mode* is one-dimensional, i.e., the vibration begins in the bond connecting the two atoms when they get excited beyond 600 K. For monoatomic substance, no vibrational mode can be seen, but there is translation mode and rotational mode to store the internal energy.

Second law of thermodynamics

It defines the entropy of the system. This law states that heat can never travel from colder region to hotter region without an external aid, i.e., $dS = dQ/T$, where dS is the change in entropy, dQ is the change in heat and T is the temperature.

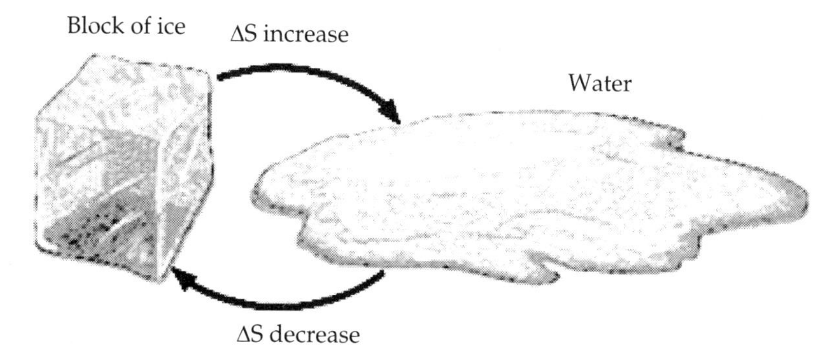

Figure 1.4 Second law of thermodynamics.

For example, water does not flow from lower level to higher level without an external aid like a pump. As shown in Figure 1.4, the block of ice melts into liquid water at room temperature, but to convert the liquid water into a block of ice, we need to cool it using some system, which is not natural. Entropy is nothing, but the energy which is neither available to do useful work nor a loss or internal energy. In a process where system transforms from one thermodynamic state to another, entropy comes into the picture. The entropy of a system remains same if the process is reversible, which is not natural, and entropy increases if the process is irreversible. The entropy of the universe (where it is irreversible process) keeps on increasing. Rust is the best example of entropy. Another example is refrigerator, which reduces the temperature of the object kept inside it, but in doing so, as some work is done; so, some amount of heat is liberated to the atmosphere. This heat is called *entropy*. The fan blows air, but in doing so, its component (electric motor) gets heated up. This heat is also called *entropy*. Equation (1.3) describes the first law of thermodynamics and the heat supplied is written further in terms of entropy, i.e., Eqs. (1.4) and (1.5).

$$dQ = dU + dW \tag{1.3}$$

$$TdS = dU + PdV \tag{1.4}$$

$$TdS = C_V dT + PdV \tag{1.5}$$

where, T is the temperature, ds is the change in entropy, P is the pressure, dV is the change in volume and C_V is the specific heat value at constant volume.

Third law of thermodynamics

The entropy of a pure substance in complete thermodynamic equilibrium becomes zero at the absolute zero temperature.

Specific heat at constant pressure

It is the amount of heat added to the system to raise the temperature of a system by a unit degree at constant pressure process. It is denoted by C_P and its unit is J/kgK.

Specific heat at constant volume

It is the amount of heat supplied to the system to raise the temperature of the system by a unit degree at constant volume process. It is denoted by C_V and its unit J/kgK.

The specific heat at constant pressure is always greater than the specific heat at constant volume because in constant volume process, there is no work done; whatever heat is supplied is converted into internal energy. Hence, it makes a significant difference between specific heat at constant pressure and specific heat at constant volume in the T-S diagram. The ratio of specific heat at constant pressure to the specific heat at constant volume is called *specific heat ratio* (γ); i.e.,

$$\gamma = \frac{C_P}{C_V} \quad \text{or} \quad \gamma = \frac{n+2}{n} \tag{1.6}$$

where n is the degree of freedom. For air (monoatomic gas), it is $n = 5$ and for combustion gas or carbon dioxide (diatomic gas), it is $n = 6$.

$$C_P = \frac{\gamma R}{\gamma - 1} \qquad C_V = \frac{R}{\gamma - 1} \qquad R = C_P - C_V$$

where R is the gas constant.

Enthalpy of the system

It is the total heat content in the open system at constant pressure process. It is denoted by H. The enthalpy of a system can also be defined as the sum of the change in internal energy and work done at constant pressure process, as shown in Eq. (1.7) and (1.8). The enthalpy variation on heat exchange is shown in Figure 1.5.

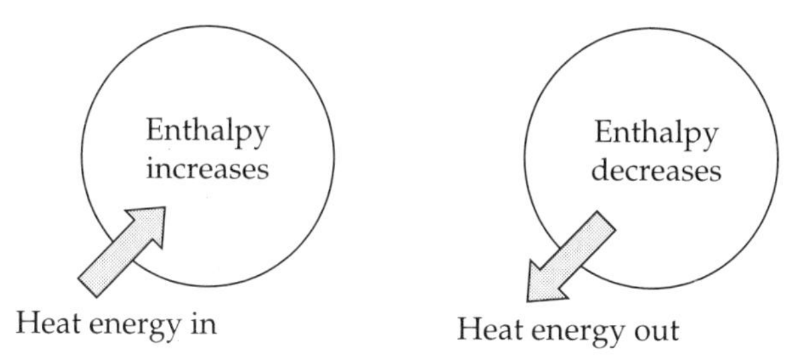

Figure 1.5 Enthalpy of a system.

$$H = U + PV \tag{1.7}$$

$$h = u + \frac{P}{\rho} = C_V T + RT = T(C_v + R) = C_P T \tag{1.8}$$

where, h is the specific enthalpy, ρ is the density, u is the specific internal energy and P is pressure.

Enthalpy is positive for an endothermic reaction and it is negative for an exothermic reaction. The power output of gas turbine components can be expressed in terms of enthalpy and used in defining efficiency too. For example, the compressor increases the pressure of the air and this increase in pressure leads to increase in temperature, and hence, the enthalpy increases. For a given mass flow rate of air, the change in enthalpy is nothing, but power.

1.2 NEWTON'S LAWS OF MOTION

The aircraft propulsion system gives momentum to an aircraft on the ground as well as in the air. The change in momentum of aircraft is because of the reaction force which is nothing, but the thrust generated in engine (residential inertia force of motion is considered if aircraft is already in motion). Hence, it is very important to know the theory behind the force produced to get the momentum. Newton's laws of motion help in knowing, understanding and applying their concepts to calculate or design the force to be generated by these power plants or propulsive devices very easily. The laws are discussed here.

Newton's first law of motion

It is also known as the *law of inertia*. The body will remain in a state of rest or motion until and unless an external force is applied to it. For example, consider you are travelling in a vehicle with certain velocity or speed and suddenly if brakes are applied, then you will get the forward movement. This phenomenon is called *inertia,* as shown in Figure 1.6.

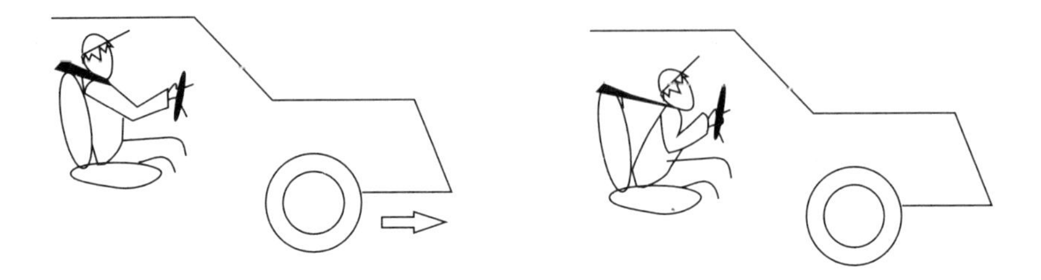

Figure 1.6 Law of inertia.

At steady level flight condition, the aircraft remains in the state of motion as other forces (mainly drag, etc.) are balanced using thrust, and also, if the vector sum of all the forces acting on an object is zero, then the velocity or momentum of the object is constant.

Newton's second law of motion

The acceleration of a body is directly proportional to the force acting on a body in the same direction and inversely proportional to the body mass. It can also be defined as the rate of change of the momentum is directly proportional to the force applied on the body, i.e.,

$$F = ma \quad \text{or} \quad F = \dot{m}v$$

where, F is the force, m is the mass, a is the acceleration, v is the velocity and $\dot{m}$ is the mass flow rate.

The unit of the force is Newton (N) or sometimes kilogram (kg) is also used. For example, to make the aircraft propel forward, the engine must expel or throw the amount of gas at a particular velocity.

Newton's third law of motion

It states that for every action there is an equal and opposite reaction. For example, consider a rocket generating exhaust jet from its engine in a downward direction, which is an action, and because of that jet, the rocket moves upward, which is a reaction. Also a man fires a gun in one direction, which is an action, and he feels reaction force from the gun.

1.3 MODES OF HEAT TRANSFER

Three modes of heat transfer are as follows:

1. Conduction: It is a mode of heat transfer where the heat transfer takes place from one end to another end of the same system or different systems without any molecular movement. For example, heating a metal rod from one end, and after some time, heat is felt at another end of the metal rod.

2. Convection: It is a mode of heat transfer where the heat is transferred from one end to another end of the same system or different systems with the molecular movement. For example, when hot water is made to flow in a metal pipe, the pipe also gets heated. If two metal pipes, one inside another, have two different fluids at a different temperature, there will be heat exchange between the fluids.

3. Radiation: It is a mode of heat transfer without any medium or external aid. For example, body feels the heat of solar radiation.

1.4 THRUST

Thrust is a force generated or force produced by an aero engine by pushing the gas out from the engine at high velocity and pressure, which moves the aircraft from its rest or initial position. This is called *forward thrust*. The same force can be used to stop the aircraft by directing the gas in opposite direction, which is generally called *thrust reversal* or *reverse thrust*.

How is thrust generated?

Thrust is a force generated by the reaction of the gas, which is produced in the combustion chamber and expanded or accelerated in the nozzle at high-velocity jet. The high-velocity jet from the nozzle creates the momentum to the flight. Also, by pushing the air using propellers/ducted fans or unducted fans without any contact of combustor or any turbomachines, thrust can be produced.

When the high temperature gas or working fluid is accelerated out from the aircraft engine in backward direction, it results into aircraft acceleration in the forward direction. The action of the fluid from the nozzle leads to reaction force on the aircraft that is thrust.

$$F = ma = \dot{m}(v_e - v_a) + A(P_e - P_0) \tag{1.9}$$

where, F is the thrust, m is mass, a is acceleration, $\dot{m}$ is mass flow rate, v_e is the exit jet velocity and v_a is the vehicle velocity, A is the exit cross-sectional area of engine, P_e is the exit pressure and P_0 is the ambient pressure.

Equation (1.9) describes that the thrust generated by the engine is the sum of momentum thrust and pressure thrust. The momentum thrust is the product of mass flow rate of burnt gases (fuel and air mixture mass flow rate) and the difference between the jet velocity and flight velocity. This thrust is called *momentum thrust*. The other type of thrust is *pressure thrust*, which is the product of exit area of engine and the difference in the nozzle exit pressure and ambient pressure.

The mechanism of thrust creation can be explained by simple fluid statics. When the high pressurised air is at the junction between the compressor and combustion chamber, then it exerts forward motion on the compressor. The high pressure air at the junction between the combustion chamber and turbine creates forward motion on the combustion chamber. As in turbine and nozzles, the pressure drops, which leads to backward motion of turbine as well as a nozzle. The thrust generated in the jet engine has both forward thrust and backward thrust. The net thrust is a combination of forward thrust and backward thrust.

1.5 EFFICIENCY

The *efficiency* of a system can be generally defined as the ratio of the output of the system to the input to the system. The input and output quantity is nothing, but the power or energy for a given interval.

Thermal efficiency, η_{Th} is the ratio of power output from the engine (thrust power) to the heat supplied to the engine (fuel energy), as expressed in Eq. (1.10). It can also be defined as the ratio of the net work done by the turbomachines at a given interval to the fuel consumed. The heat supplied can be written as the product of mass flow rate of the fuel and its calorific value.

$$\eta_{Th} = \frac{\dot{m}v_a(v_e - v_a)}{\dot{m}C_V} \tag{1.10}$$

Transmission efficiency, η_{Tr}, is the efficiency of a shaft or spool when the energy gets transferred through the shaft connecting turbine and propeller through gearbox as expressed in Eq. (1.11). This efficiency can be seen only in turboprop and turboshaft engines, where a shaft is extended out from the engine which drives the propeller. The input from the turbine to the shaft is huge; hence, it is required to use a reduction gearbox, which reduces the speed of the rotating shaft. The output from the reduction gearbox and the input from the turbine is used to find the transmission efficiency.

Mechanical efficiency is also a type of efficiency, which is related to only turbomachines, i.e., it is the ratio of work done by the compressor to the work done on the turbine.

$$\eta_{Tr} = \frac{\text{Transmission output}}{\text{Transmission input}} \tag{1.11}$$

Propulsive efficiency, η_P, is the ratio of thrust power to the sum of thrust power and the initial kinetic energy, as expressed in Eqs. (1.12) and (1.13). Propulsive efficiency is maximum when the exit jet velocity from the engine is equal to the inlet air velocity or vehicle velocity. The propulsive efficiency decreases with a decrease in the vehicle velocity.

$$\eta_P = \frac{\dot{m}v_a(v_e - v_a)}{\dot{m}\left[v_a(v_e - v_a) + \frac{(v_e - v_a)^2}{2}\right]} \quad (1.12)$$

$$\eta_P = \frac{2v_a}{v_e + v_a} = \frac{2\alpha}{1+\alpha} \quad (1.13)$$

where, α is the ratio of vehicle velocity to the exit jet velocity.

Overall efficiency, η_0, is the product of thermal efficiency, transmission efficiency and propulsive/propeller efficiency, as expressed in Eq. (1.14).

$$\eta_O = \frac{\text{Engine output}}{\text{Engine input}} \times \frac{\text{Transmission output}}{\text{Transmission input}} \times \frac{\text{Propeller output}}{\text{propeller input}}$$

$$\eta_O = \eta_{Th} \times \eta_{Tr} \times \eta_P \quad (1.14)$$

1.6 BRAYTON CYCLE

Brayton cycle is the thermodynamic cycle for all types of gas turbine engines and other propulsive systems like ramjet engines. In this cycle heat is added to the system when the system is in constant pressure process; hence, it is also named as *constant pressure heat addition cycle.*

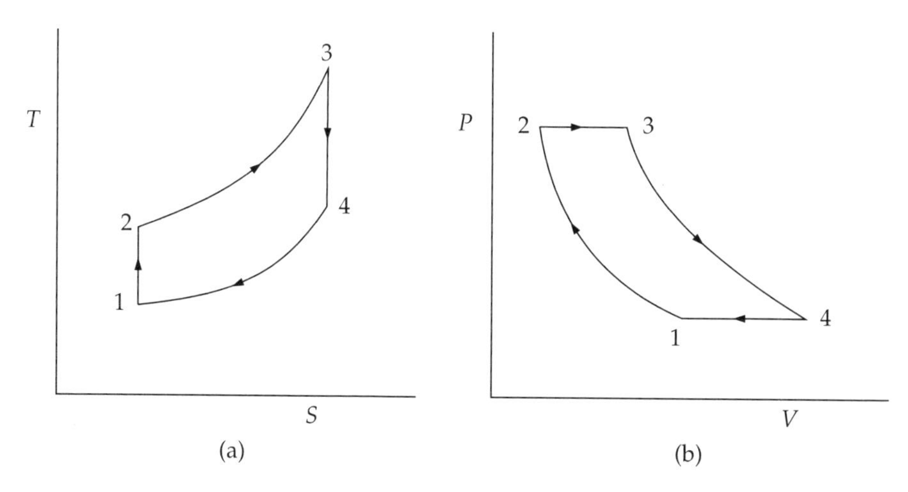

Figure 1.7 (a) *P–V* and (b) *T–S* diagram for Brayton cycle.

As shown in Figure 1.7, the Brayton cycle has four basic processes. Process 1–2 is isentropic compression of air in intake, fan and compressor sections. In this process, air pressure and temperature increases at constant entropy. Process 2–3 is constant pressure heat addition process in combustion chamber or burner, where the thermochemical reaction of fuel and oxidant takes place with no change in the pressure inside the combustion chamber. Process 3–4 is isentropic expansion of air in turbine and nozzle sections. In this process, the pressure energy is converted into kinetic energy and the temperature of fluid also decreases at constant entropy. Process 4–1 is exhaust or heat removal process by exhausting the hot fluid to atmosphere where the pressure and temperature reduces to atmospheric level.

1.7 HUMPHREY CYCLE

Humphrey cycle is the thermodynamic cycle which is the reference for few engines like pulse jet engines and pulse detonation engine. In this cycle, heat is added to the system when the system is in constant volume process; hence, it is also named as constant volume heat addition cycle.

As shown in Figure 1.8, it has four processes. Process 1–2 is isentropic compression, which means the air or fluid is compressed with no change in the entropy. Process 2–3 is constant volume heat addition in the combustion chamber or heat addition process, which means the fuel and air combustion takes place at constant volume. Process 3–4 is isentropic expansion, and finally, process 4–1 is exhaust or heat removal process.

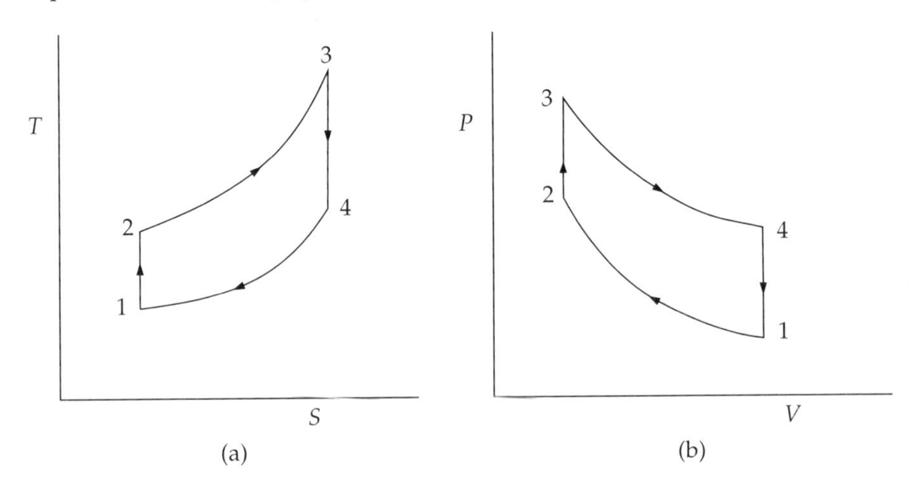

Figure 1.8 *P–V* and *T–S* diagram for Humphrey cycle.

1.8 GOVERNING EQUATIONS

To understand the propulsion technology, it is also necessary to study the governing equations, which explain the mass, momentum and energy available inside the system. These governing equations are used to define and relate the parameters like thrust, internal energy, enthalpy, work done and heat supplied to the system. These are also used to find the stagnation condition parameters.

The governing equations also give the difference between the actual and ideal condition of a process. These equations can also be used for deriving the advance mathematical relations like Mach number relations, area Mach number relations, and many more.

Continuity equation

This equation describes that the mass flow is conserved inside the system. It means the amount of fluid flow entering the system is the same amount of fluid flow leaving the system, as shown in Figure 1.9.

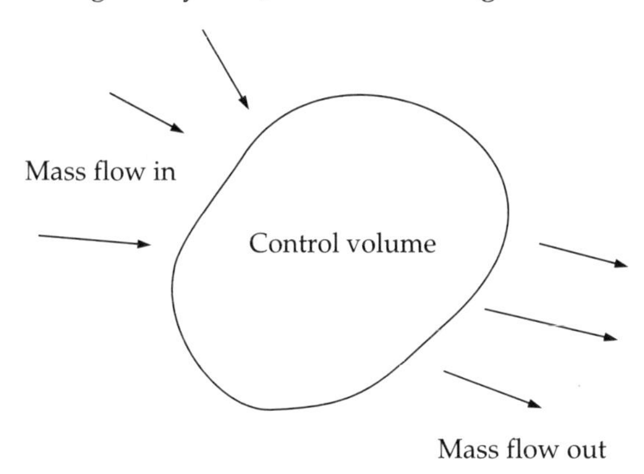

Figure 1.9 Conservation of mass.

This equation or condition is used to relate the properties of fluids which are entering and leaving the system, as shown in Eq. (1.15), where ρ is density (kg/m^3), A is the area (m^2), v is the velocity (m/s), subscript 1 defines input condition and subscript 2 defines output condition.

$$\rho_1 A_1 v_1 = \rho_2 A_2 v_2 \tag{1.15}$$

Momentum equation

This equation describes that the momentum of fluid is conserved, i.e., Eq. (1.16), where P is the pressure. The change in momentum of a fluid through the system is constant. The momentum force is expressed in terms of the fluid velocity and fluid pressure.

$$P_1 + \rho_1 v_1^2 = P_2 + \rho_2 v_2^2 \tag{1.16}$$

Energy equation

This equation describes that the energy is conserved. The energy of the flow which is entering the system or consumed by the system is same as the energy of the flow which is leaving the system or produced by the system. This energy is the sum of internal energy, kinetic energy and potential energy, as shown in Eq. (1.17). The change in energy is the sum of the change in internal energy, change in kinetic energy and change in potential energy as, shown in Eq. (1.18).

$$E = U + mgl + \frac{1}{2} mv^2 \tag{1.17}$$

where, E is the total energy, U is internal energy, g is acceleration due to gravity and l is height.

$$dE = dU + mgdl + \frac{1}{2} mdv^2$$

$$E_2 - E_1 = U_2 - U_1 + mg(l_2 - l_1) + \frac{1}{2} m(v_2^2 - v_1^2) \tag{1.18}$$

As the thermodynamic first law describes, the heat energy supplied is transformed into work done and some other form of energy, as shown in Eq. (1.19).

$$Q = W + \Delta E \tag{1.19}$$

where, Q is heat supplied and W is work done.

$$Q = W + (U_2 - U_1) + mg(l_2 - l_1) + \frac{1}{2} m(v_2^2 - v_1^2)$$

$$q = w + (u_2 - u_1) + g(l_2 - l) + \frac{1}{2}(v_2^2 - v_1^2) \tag{1.20}$$

where, q is specific heat supplied, is specific internal energy and w is specific work done.

Equation (1.20) is the energy equation. The total work done is the sum of shaft work done and thermodynamic work done. The same concept is reflected in Eq. (1.21).

$$Q - (W_s + P\Delta V) = m(\Delta u + \Delta KE + \Delta PE) \tag{1.21}$$

where, W_s is the shaft work, KE is the kinetic energy and PE is the potential energy.

$$Q = W_s + (P_2V_2 - P_1V_1) + (U_2 - U_1) + mg(l_2 - l) + \frac{1}{2} m(v_2^2 - v_1^2) \tag{1.22}$$

$$Q = W_s + (H_2 - H_1) + mg(l_2 - l_1) + \frac{1}{2} m(v_2^2 - v_1^2)$$

where H is enthalpy.

$$Q + H_1 + mgl_1 + \frac{1}{2} mv_1^2 = W_s + H_2 + mgl_2 + \frac{1}{2} mv_2^2 \tag{1.23}$$

$$q + h_1 + gl_1 + \frac{1}{2} v_1^2 = w_s + h_2 + gl_2 + \frac{1}{2} v_2^2 \tag{1.24}$$

where, w_s is specific shaft work, and h is specific enthalpy.

Equation (1.23) is called *steady flow energy equation,* which explains about the heat, work and other energy transaction in major gas turbine components. The equation links the various thermodynamic properties, work done with energy terms of a gas turbine.

Stagnation conditions

This is a condition of the flow where the flow is adiabatically decelerated to zero velocity, represented by subscript zero.

Stagnation enthalpy

It is the enthalpy of the flow when the flow is adiabatically decelerated to zero velocity at zero elevation. In Eq. (1.25), h_0 is the stagnation enthalpy, h_1 is the initial enthalpy before flow deceleration and v_1 is the initial velocity before deceleration.

$$h_0 = h_1 + \frac{1}{2} v_1^2 \tag{1.25}$$

Stagnation temperature

It is the temperature of the flow when the flow is adiabatically decelerated to zero velocity at zero elevation.

$$h_0 = h_1 + \frac{1}{2} v_1^2$$

$$C_P T_0 = C_P T_1 + \frac{1}{2} v_1^2 \tag{1.26}$$

The stagnation enthalpy can be written in terms of specific heat and temperature, as shown in the above Eq. (1.26).

$$T_0 = T_1 + \frac{1}{2C_P} v_1^2$$

where T_0 is the stagnation temperature, and T_1 is the initial temperature before stagnation condition.

In the stagnation temperature equation, the specific heat at constant pressure can be written in the form of specific heat ratio, as shown in the Eqs. (1.27) and (1.28).

$$T_0 = T_1 + \frac{(\gamma - 1)}{2\gamma R} v_1^2 \tag{1.27}$$

$$T_0 = T_1 \left(1 + \frac{(\gamma - 1)}{2a^2} v_1^2\right)$$

$$T_0 = T_1 \left(1 + \frac{(\gamma - 1)}{2} M_1^2\right) \tag{1.28}$$

where, γ is the specific heat ratio and M is the Mach number.

Stagnation pressure

It is the pressure of the flow when the flow is adiabatically decelerated to zero velocity with zero elevation. In Eq. (1.29), P_0 is stagnation pressure and P_1 is the initial pressure before stagnation.

$$P_0 = P_1 \left(1 + \frac{(\gamma - 1)}{2} M^2\right)^{\frac{\gamma}{(\gamma - 1)}} \tag{1.29}$$

1.9 CLASSIFICATION OF PROPULSION SYSTEM

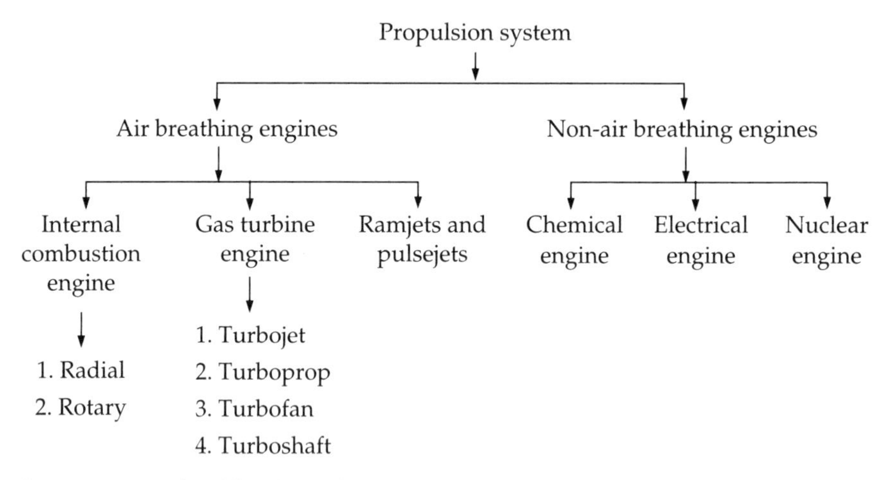

Figure 1.10 Classification of propulsion system.

Aero propulsion system is mainly classified into two major groups, namely, air breathing engines and non-air breathing engines (Figure 1.10). In simple words, *air breathing engines* are those engines which take or consume air from the atmosphere for the combustion process, whereas the *non-air breathing engines* are those which do not consume air from the atmosphere for the combustion process. The meaning of consuming air from atmosphere is that the engine sucks in the atmospheric air through the inlet and it is used for combustion and other processes.

Air breathing engines are again classified into three main categories—internal combustion engine, gas turbine engine and ramjet and pulse jet engines. In many aircrafts, internal combustion engine is used as a rotary engine and radial engine by having multiple internal combustion engine blocks together in a circular arrangement to drive the propeller. Initially, IC engines were used as multiple internal combustion engine blocks in a series arrangement to drive the propeller, but later, these engines were used as radial and rotary engine to reduce the engine length.

1.10 INTERNAL COMBUSTION ENGINE

The engine in which combustion takes place within the cylinder block or in a closed system without any momentum of the charge and the heat energy is transformed into linear mechanical motion is called *internal combustion engine.*

This is an engine that uses one or more reciprocating pistons and cylinder arrangement to convert pressure and heat energy into linear motion or mechanical form of energy. The crack shaft is a mechanical device, which converts the linear motion of piston using connecting rod into rotary motion of the power shaft or torque tube with or without a flywheel. A flywheel is often used to ensure smooth rotation of the power shaft and also to store energy. This engine can operate to have a flight velocity of 250 mph at maximum condition. For low altitude, simple engine can be used, but for high altitude, turbo-supercharged engines are used.

Working cycles of IC engine

1. Intake stroke: The charge of calculated amount is taken into the combustion cylinder.

2. Compression stroke: The charge is compressed and pressurised by moving the piston from bottom dead centre to top dead centre of the cylinder.

3. Power stroke: The combustion process is initiated by producing spark using a spark plug electrode and combusting the charge.

4. Exhaust stroke: Combusted products are expelled out from the engine.

Requirements of IC engine

1. Sparking must occur at proper time in all the conditions.
2. Engine must be vibration-free.
3. Fuel metering device should be proper.
4. Oil system must deliver at proper pressure.

The ignition sparking in IC engine should be at the right time without any delay or pre-ignition, which leads to drop in efficiency and more fuel consumption. The ignition source should be working properly in all weather conditions without any cold problem. In conversion or transformation from heat energy to mechanical energy, the vibration should be very less. The vibration leads to the structural problem and noise.

The fuel measuring device should be proper which must show the proper fuel consumption. If the fuel metering device does not show the correct consumption, then the prediction for future fuel consumption and duration of flight will be very difficult. The reciprocating engine requires twice the instrumentation as compared to the gas turbine engine. Any change in power setting of IC engine leads to at least five control adjustments.

The *compression ratio* of reciprocating engine or IC engine is the ratio of space inside the cylinder when the piston is at bottom dead centre (BDC) to space inside the cylinder when the piston is at top dead centre (TDC) as shown in Figure 1.11. Engine with high compression ratio is used for long range fuel economy and to avoid knocking due to manifold increase in pressure as well as temperature from the supercharger. The compression ratio decides the power output of the engine.

Efficiency of the IC engine is the ratio of thrust horsepower to the brake horsepower of the engine.

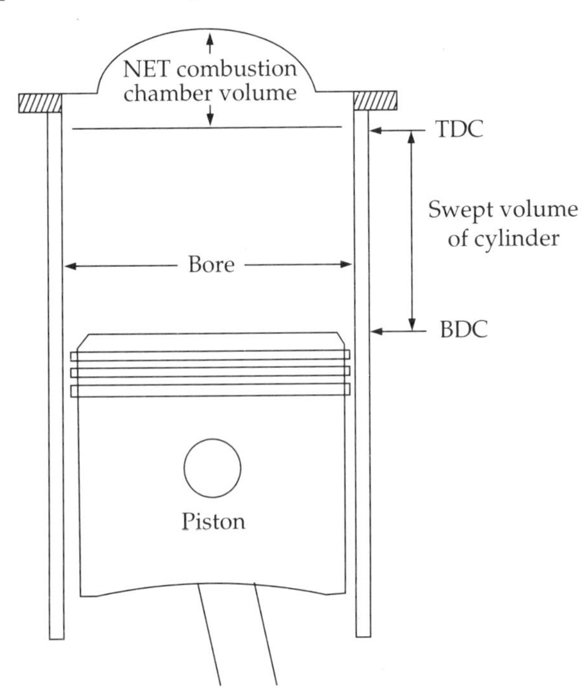

Figure 1.11 Dimensional parameters of internal combustion engine.

The internal combustion engine drives the propeller, which generates forward thrust. During take-off, the blade angle of the propeller blade is acute angle, which gives maximum thrust, whereas during cruise, thrust requirement will be less; so, the blade angle of the propeller is adjusted to the higher side of the acute angle. The engine propellers are of three types, namely, fixed pitch, reversible pitch and controllable pitch.

In *fixed pitch propellers*, the blades cannot be moved or adjusted during flight and these will give optimum efficiency at specific speeds. *Reversible pitch propellers* are those where the propeller blades' angle of attack can be reversed for shorter landing and ground maneuvering, which means the blade angle of the propeller goes beyond right angle or it is an obtuse angle. *Controllable pitch propellers* are those where the blade angle can be varied during flight based on the requirement of thrust.

The main components of internal combustion engine are cylinder block, cylinder head, piston, connecting rod, piston rings, inlet and exit ports or ports, crankshaft, crankcase and igniter. The IC engine inlet and exhaust valves are of two types, namely, mushroom type and tulip type, which are usually made up of chromium-nickel steel. The cylinder head or cylinder block is the space where the piston reciprocates, which is made up of aluminum alloy whereas valve seats are made up of steel, tin-bronze and aluminum-bronze. To hold the valves firmly on their position, spring power is used and that is why it is called *valve spring*.

The ignition or firing order for multi-cylinder IC engine varies based on the number of cylinders and alignment of cylinders, as shown in Figure 1.12. Firing order for single row 9-cylinder is 1-3-5-7-9-2-4-6-8 [Figure 1.12(a)], whereas firing order for double row 18-cylinder is 1-12-5-16-9-2-13-6-17-10-3-14-7-18-11-4-15-8 [Figure 1.12(b)] and firing order for double row 14-cylinder is 1-10-5-14-9-4-13-8-3-12-7-2-11-6 [Figure 1.12(c)]. Firing order is nothing, but the initiation of power stroke using the sharp sparking in the individual engine block.

The firing order follows this pattern because these engines are connected to the crankshaft and the energy should transfer to the crankshaft smoothly with less loss (vibration-free). Firing order is made to make the maximum utilisation of power produced by the series of internal combustion engine and to design the common shaft, i.e., crankshaft, which transforms the reciprocating motion into rotary motion. Heating of engine block due to combustion process is rapid; hence, care must be taken to avoid some problems like feathering of piston rings, scored cylinders, burned valves and piston heads.

Engine faults or problems can be detected through the following steps:

1. Check the sparking at the ignition plugs, and if there is discontinuous sparking then rectify and solve the route causes like carbon deposition or mag wire grounding or defective switch.

2. If the spark is weak then the engine will not start and the reason for weak spark is dirty or contaminated igniter ends.
3. Insufficient fuel supply also leads to engine problem, which may be because of insufficient priming or defective fuel pump or vaporization of fuel in the line or due to closing the throttle valve.
4. Excessive priming or high pressure leads to flooding.
5. Defective magneto also results in the incorrect timing of system which leads to faulty results.

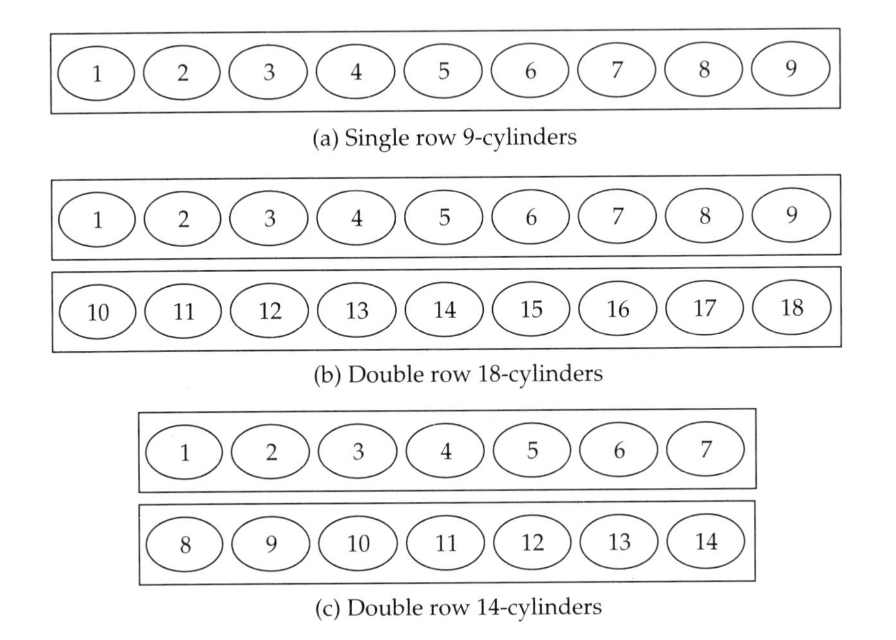

Figure 1.12 Ignition order for multi-cylinder engines.

1.10.1 Turbo-charged Internal Combustion Engine

A *turbocharger* is an additional augmenting device attached to the IC engine, which is used to increase the engine performance and power output. As shown in Figure 1.13, it has a set of compressor (C) and turbine (T) mounted on the same shaft (spool). Exhaust gas from engine block is made to impinge on the turbine blades, which drives the turbine, and hence, compressor is connected to the turbine. The rotating compressor sucks the air from the atmosphere and compresses it to high pressure. The high-pressure air is supplied to the engine, which enhances the performance of the engine, as shown in Figure 1.13.

If power requirement is less, then the waste gate is opened, which is located in between the exhaust pipe and the turbo-charger. This leads to directing the exhaust gases away from turbine line and also, it slows down the RPM (revolution per minute) of the turbine, and hence, the compressor RPM.

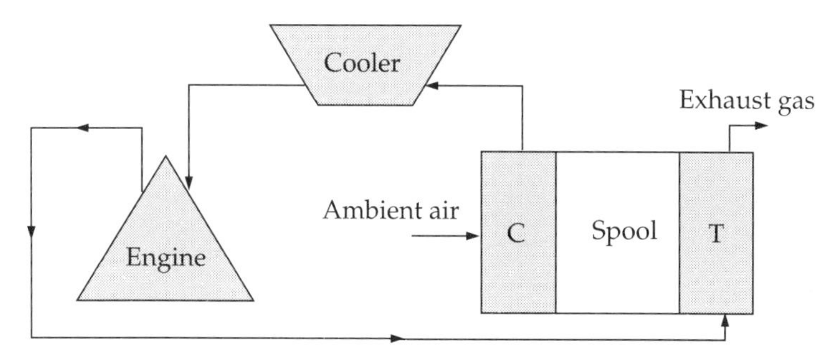

Figure 1.13 Turbo-charged engine.

1.11 GAS TURBINE ENGINE

As the name indicates, the *gas turbine engine* is one where the working fluid (air) is continuously taken inside the engine and expelled out at high momentum or just high velocity. The air is used to combust the fuel with the help of ignition source in the combustion chamber and that combusted gas is made to impinge on the turbine, thus making the turbine to rotate, which in turn drives the compressor, and the high-velocity jet is produced at the exit section of the engine. To have the most efficient combustion, the input energy (which is air and fuel) should be better; hence, the air is pressurised before combustion using a set of compressors, which are driven by the turbine.

As shown in Figure 1.14, the leading part of a basic type of gas turbine engine is intake followed by compressor, combustion chamber, turbine, and finally, exhaust duct or nozzle. The shape of the inlet looks like a divergent duct, which acts as decelerating duct and helps the compressor in transforming the kinetic energy into pressure energy. The pressurisation of air is done using the compressor, which is located after inlet or fan or propeller of a particular type of gas turbine engine.

The compressed air is supplied to the combustion chamber to have the thermo-chemical reaction of fuel and air mixture. The hot gas generated in the combustion chamber is directed to impinge on turbine blades using guide vanes. The combusted gas gives the angular movement to the turbine blades, which is nothing but a rotary motion to the turbine. The rotation of turbine leads to some mechanical work such as driving the compressors, electric power generators and pumps, etc. The hot gas expanded in the turbine is made to flow through the nozzle, which also accelerates the flow by transforming the remaining pressure energy into kinetic energy. The expansion means the pressurised gas gets transformed into high-velocity gas.

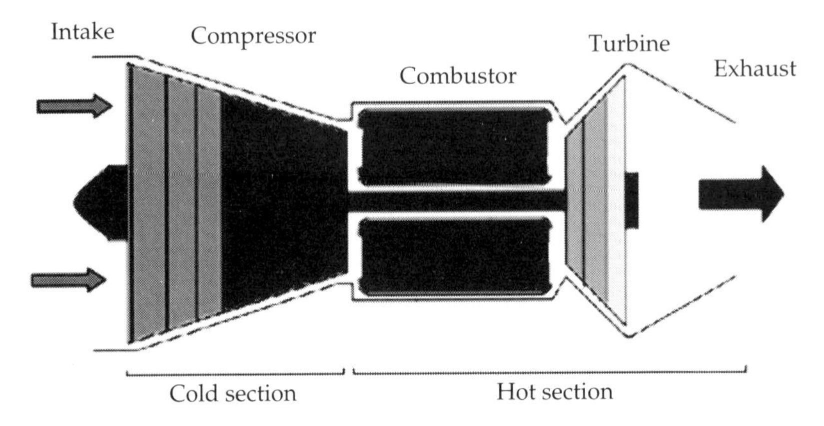

Figure 1.14 Gas turbine engine.

The performance and efficiency of the engine depend on the engine pressure ratio and specific fuel consumption. *Engine pressure ratio* is the ratio of turbine discharge product pressure to the engine inlet air pressure. The pressure ratio helps in producing momentum, and hence, thrust. *Specific fuel consumption* is the amount of fuel consumed to generate a unit thrust or inverse of specific impulse. *Specific thrust* is the amount of thrust generated using a unit mixture of fuel and air. As the fuel consumption increases, the efficiency drops down.

As shown in Figure 1.15, the inlet air temperature of low-pressure compressor (LPC) is up to 70°C, and input to the high-pressure compressor (HPC) is around 200°C. The inlet air temperature to the combustion chamber (CC) is 420°C (maximum) and the combustion temperature inside the combustion chamber can rise up to 2400°C. The input to the high-pressure turbine (HPT) ranges from 1000°C to 1600°C. The input to low-pressure turbine (LPT) is around 800°C and exhaust nozzle temperature is almost 700°C.

Air entering the inlet is compressed by the compressor, which has multi-stages. Sometimes, water or alcohol is sprayed in the compressor to increase its performance by making the air denser. The fuel is injected and burned in the combustor annular space between the outer shell and the spool cover or inner shell. The hot pressurised combusted gas then drives the turbine disk and gets expanded. The hot exhaust gases expand more through the nozzle to match with ambient pressure and produce thrust.

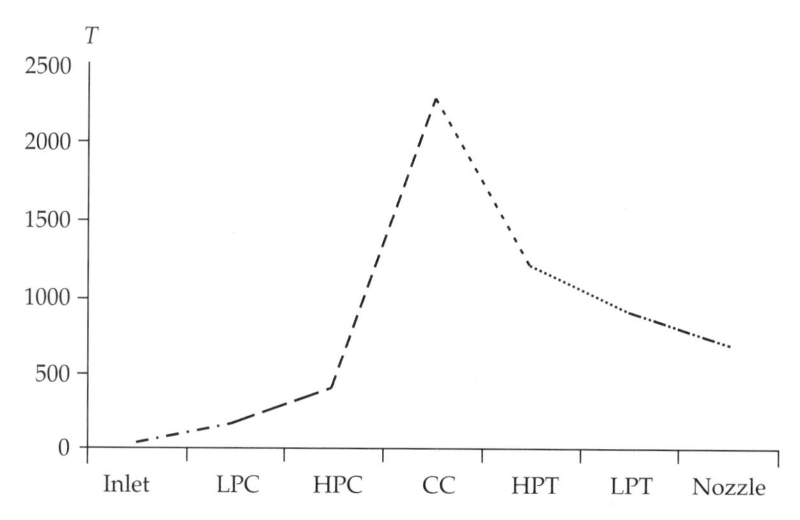

Figure 1.15 Temperature variation in gas turbine engine.

1.12 RAMJET ENGINE

Ramjet engine is also known as *flying stove*. The name 'ramjet' is because of the method used for air compression. In this engine, the air is compressed by ramming the air. Ramming is done because of the aerodynamic shape of the inlet and vehicle velocity. Ramjet engine cannot operate from a stagnant point or zero velocity, as the air must enter the engine without external work like compressor work. It must have some amount of acceleration to high speed by other means of propulsion. This engine does not have any rotating parts.

As shown in Figure 1.16, the air enters the spike-shaped inlet, where the flow gets accelerated negatively, and the diffuser which is next to inlet acts as a compressor. In this, the supersonic air creates shock waves, which are reflected in between the upper and lower engine inlet lips or cowling and the spike. The shock reflection leads to changing the physical properties of the flow stream entering the engine across the shock. The physical properties are nothing, but the velocity, density and pressure of the flow.

At the end of the spike, the flow pressure will be high and velocity will be low enough for the combustion process. The propellant is injected into the combustor by some means and combustion is carried out in the presence of ignition source. Flame holders are used to stabilise the flame from the high-velocity incoming air. The burning of fuel imparts thermal energy to the gas, and the expansion through the nozzle at high speed greater than the incoming air speed or vehicle speed produces the forward thrust.

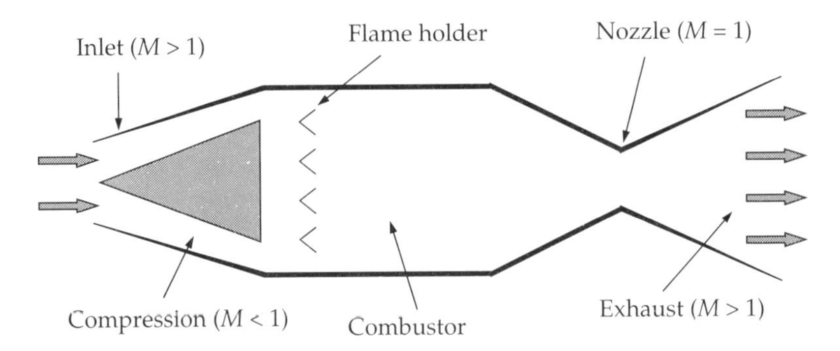

Figure 1.16 Ram Jet Engine.

1.13 ROCKET ENGINE

A typical liquid propellant rocket consists of a guidance compartment, payload, fuel tank, oxidant tank and engine, as shown in Figure 1.17. A gas generator operated from the main propellant or an auxiliary propellant drives to supply fuel and oxygen at a pressure of 20–40 times the atmospheric pressure to the thrust chamber. The fuel is usually circulated in a cooling jacket surrounding the nozzle and combustion chamber prior to injection and burning, which is called *Regenerative cooling,* to maintain the surface temperature of nozzle and combustor at a value lower than the critical temperature limit and to inject preheated fuel into the combustion chamber for better combustion performance.

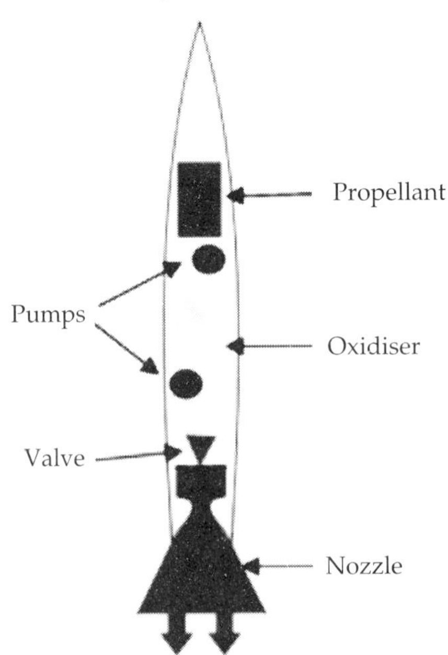

Figure 1.17 Rocket engine.

The fuel and oxidiser combust in the combustion chamber and hot gases expand through the nozzle to produce forward thrust. In rockets, impulsive thrust is produced as air or oxidiser is not taken from the atmosphere.

1.14 FACTORS AFFECTING THE THRUST

The following factors affect the thrust:

1. Nozzle jet velocity: The velocity of the gas coming out from the engine or nozzle is called *nozzle jet velocity*. As per the law of conservation of momentum or Newton's second law, the increase in exit velocity increases the thrust value and decreases with decrease in the exit velocity.

2. Air speed: When aircraft is moving, the air velocity v_1 (vehicle velocity) entering the engine increases. This leads to less change in the momentum thrust. But, if the aircraft is at rest, then the change in the momentum thrust is more. Therefore, as air speed increases, the net thrust decreases and propulsive efficiency increases.

3. Mass air flow: The most significant variable in the thrust equation is mass air flow. The air density has a profound effect on the thrust produced. The volume of air flowing through the engine is relatively fixed for any particular RPM due to the size and geometry of inlet section. Any increase in air density increases the richness of the mass flow rate, and hence, thrust. As temperature increases, its density decreases.

4. Altitude effect: It has both negative and positive effects on thrust. As altitude increases, air density decreases, which leads to decrease in the thrust because of less mass flow rate of air. Thrust increases with increase in altitude because of decrease in pressure which leads to less drag on aircraft.

5. Ram effect: The compression of air in the inlet duct or near inlet duct due to the forward motion of the vehicle is called *ram pressure* or *ram effect*. Because of the ram effect, an increase in air speed also increases the pressure and the airflow. Ram-effect depends on the shape and geometry of inlet.

6. Exit pressure: It is the pressure of the gas after nozzle or at the exit plane of the nozzle. The increase in exit pressure leads to one more component in the thrust equation called *pressure thrust*. If the exit pressure matches with the ambient, then pressure thrust value will be zero, but will get maximum thrust value.

1.15 TURBOJET ENGINE

The specifications of turbojet engine are as follows:

1. The diffuser is used initially to achieve the required pressure energy from the kinetic energy so that the momentum effect or load on the compressor is reduced.
2. Part of static pressure rise of air from its kinetic energy is done by the diffuser to reduce the compressor load.
3. The compressor can be a centrifugal type or an axial flow type or even combination of both the types, where the air is compressed to a high-pressure ratio of 9:1 to 13:1.
4. Continuous combustion is carried out in annular type combustion chamber and radial type combustor can also be used if the last stage of the compressor is the centrifugal type.

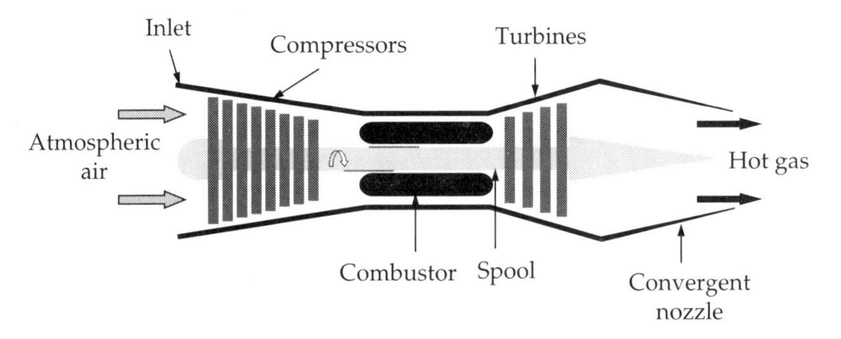

Figure 1.18 Turbojet engine.

5. Current compression ratio in the compressor is 16:1 with the combination of high-pressure set and low-pressure set.
6. Turbine inlet temperature in the order of 1200–1600 K is reduced from 2600 K by mixing with the cold air flowing around the combustor.
7. The corresponding speed of exhaust jet with aircraft propelling at 900 km/hr is of the order of 500 m/s.
8. At the downstream of the turbine, the exhaust contains some amount of unused air, which can be used for secondary combustion as thrust augmenter-afterburner.
9. Turbojets are more suitable for Mach number 1.6 to Mach number 3.5.
10. It is the most efficient engine for high altitudes.
11. The drag due to the engine is less, as the frontal area is less, hence, it can be used for maneuvering type of aircraft.

12. As the bypass air is almost negligible, the noise due to the exhaust jet is more.

1.16 TURBOPROP ENGINE

This type of engine has a propeller in front of it and a centrifugal compressor stage, which makes it simple, lightweight and less lengthy (see Figure 1.19). The efficiency of the propeller goes down when the flow velocity goes beyond 0.7 Mach number. The propeller is driven using the power turbine or free turbine which is exclusively used to drive the propeller only. The specifications of turboprop engine are given below:

1. Propeller efficiency decreases as aircraft gains the speed beyond the limit as mentioned above because the high vehicle speed leads to high-velocity air as input to the propeller and at the propeller blade tip speed will be close to supersonic velocity, which leads to shock and flow separation.
2. This engine has the lowest specific fuel combustion in the range of Mach numbers from 0.3 to 0.6.
3. Thrust to shaft power ratio for turbo prop engine is given by $W/ESHP$ = 2.9 N/kW for a range of 500 kW power engine, 2.3 N/kW for a range of 2500 kW power engine and 1.4 N/kW for a range of 7500 kW power engine.

   where, W is the thrust, $ESHP$ is the engine shaft horse power.

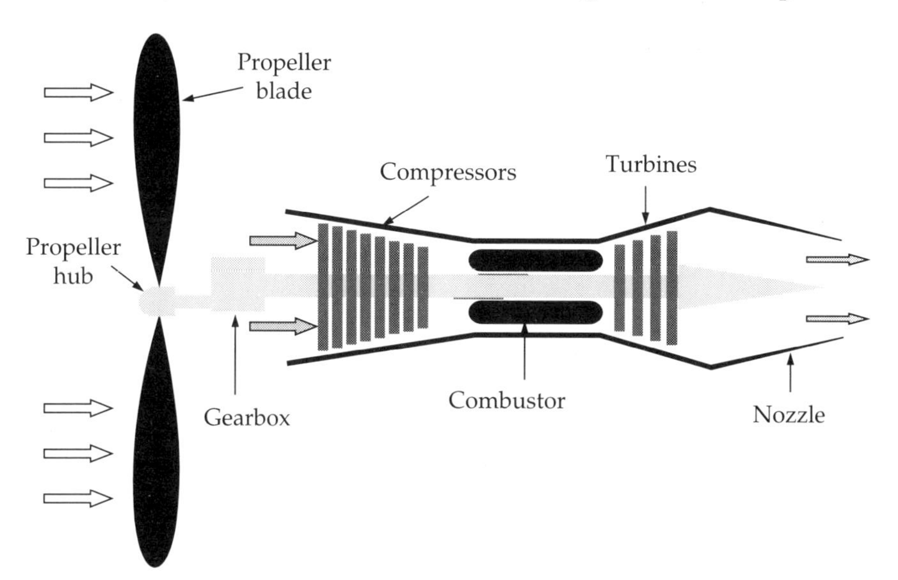

Figure 1.19 Turboprop engine.

4. It generates two types of thrust, namely, exhaust thrust and propeller thrust. The exhaust thrust is from the nozzle and the propeller thrust is from the propeller.
5. 80–90% of the total thrust is from the propeller and 10–20% of the total thrust is from the nozzle.
6. The engine can generate a flight velocity of 180 mph to 350 mph, assuming ideal case with two engines.
7. It has got low-service ceiling, which means it has a restriction to the altitude. As the altitude increases, the air becomes lighter or density goes down, and hence, more work has to be done by the propeller to push the air and produce thrust.
8. Propeller blade tip may reach a supersonic speed at high RPM, which leads to shock and boundary layer separation. Hence, at a higher altitude, to generate the same thrust, RPM can not be increased.
9. Using intermediate gearbox, turbine enables the propeller to produce controlled thrust by varying the propeller speed.

Turboprop engine propeller classification

The turboprop engine propeller is classified into the following three categories:

1. Controllable pitch: Propeller blade angle can be varied while rotating or driving the propeller assembly as per the requirement during flight.

2. Reversible pitch: The propeller blade can be made so diversified that it will generate negative thrust for short landing, roll and ground maneuvering. Such propeller blades are called *reversible pitch blades.*

3. Fixed pitch: Blade angle cannot be varied during flight which leads to an optimum efficiency of the propeller at a specific speed.

1.17 TURBOFAN ENGINE

A *turbofan engine* is a type of jet engine where a low-speed fan is used before compressor, which generates additional thrust. If the fan is mounted in front of the compressor or at the inlet stage, it is called *forward fan engine* or just *fan engine,* and if the fan is mounted after turbine section, then it is called *aft-fan engines.*

In this type of engine, a new terminology comes into a picture called *bypass ratio.* It is the ratio of the amount of air flowing across bypass duct (around the core engine) to the amount of air flowing through the core part of the engine. This happens because of the fan of very big diameter located before compressor section. The flow at the downstream of the fan is segregated into two parts. One enters the engine components and another goes around the engine. The thrust produced by the fan is called *fan thrust* and the thrust produced by core jet is called *core thrust*. The ratio of fan thrust to the core thrust is called *thrust ratio* of a turbofan engine. The ratio of air pressure at the exit of the fan to the air pressure at the entry of the fan is called *fan pressure ratio*.

The air which enters the engine components is called *primary air* (core air) and the remaining air which goes around the engine is called *bypass air* as shown in Figure 1.20. For low-bypass engine, it is 1:1, for medium-bypass engine, it ranges from 2:1 to 3:1, and for high-bypass engine, it is more than 5:1. The engine pressure ratio for low bypass engine is 15:1, and for high-bypass engine, it is more than 30:1. At present, research is going on turbofan engine to increase its efficiency and performance. The research on turbofan engine is resulting into ultrahigh bypass engine propfan and unducted fan engines. The bypass ratio in these type of engines will be enormous which leads to increase in thrust with less fuel consumption. Unducted fan engine will have bypass ratio of infinity, similar to turboprop engine. These utilise titanium lightweight stainless steel and composite materials to have better control on fuel consumption, which is more than 15%. 40:1 bypass ratio can be produced in these engines, with a fan blade diameter of 12–15 ft. It produces 10000–15000 horsepower that can carry 150–200 passengers at 0.8 Mach number. Propfan with conventional cowl type inlet can achieve 0.9 Mach number.

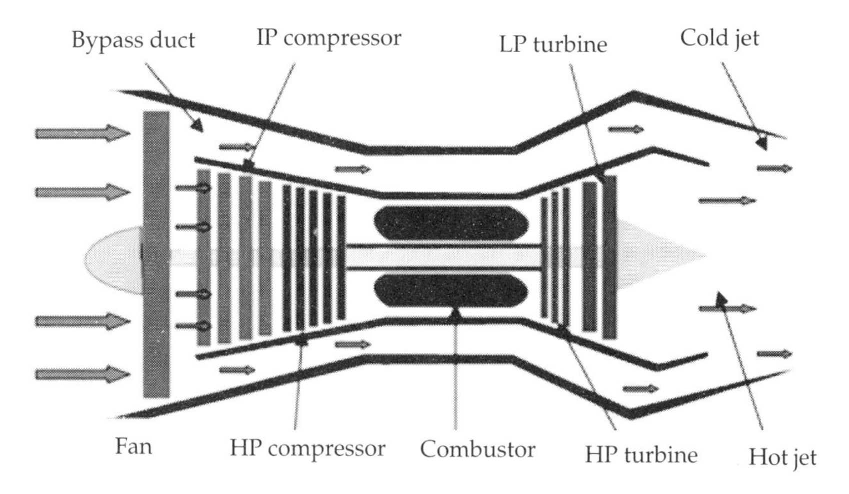

Figure 1.20 Turbo-fan engine.

The specifications of turbofan engine are as follows:

1. It overcomes the limitations of turboprop engine for high-speed application at high altitude.
2. It is the combination of turboprop and turbojet engine, where it generates the thrust from core and bypass part of the engine.
3. It has a large ducted fan, which generates two streams of air out of the engine.
4. The fan is not as large as propeller blade; hence, an increase of airspeed along the blade is comparatively less.
5. By enclosing the fan inside the cowling, the aerodynamics is better controlled.
6. The overall efficiency is very high as compared to all other gas turbine engines.
7. High bypass ratio twin spool engine has got much higher propulsive efficiency with small inner core engine thrust.
8. Ultrahigh bypass twin spool geared turbofan engine generates a high propulsive efficiency with greater bypass flow rate of air. The fan RPM is controlled using gear train.
9. A three-spool turbofan bypass engine has got a high-pressure spool, intermediate pressure spool and low-pressure spool. The low-pressure turbine drives the fan, whereas the intermediate- and high-pressure turbines drive the compressors.
10. Three spools geared contra-rotating aft-fan has got the big fan at the rear end of the engine. The complete stage of compressors is driven using only two sets of turbine. The third spool is used to drive the aft-fan using the energy from free turbine, which is next to the set of main turbines. The gearbox helps the aft-fan to rotate in opposite direction as counter-rotating type.
11. Two-spool high bypass aft-fan turbo engine has a fan mounted at its rear end on the free turbine so that transmission loss is zero and the other set of turbines drive the compressors and other devices.

1.18 THRUST AUGMENTATION

Thrust augmentation takes place through three methods, namely, afterburner, water injection and bleed air. Here, we will discuss only two methods, i.e., afterburner and water injection.

Afterburner

Afterburner is a thrust augmenting system attached at the end of the main engine to produce additional thrust. It is an extended part or additional section of an engine between turbine and exhaust nozzle. It has mainly a fuel manifold, an ignition source and a flame holder. It consists of an annular ring after turbine section, which injects some amount of fuel at the tailplane or afterburner combustor. The fuel which is injected is mixed with the hot gases coming out from the turbine and the secondary air which is flowing around the engine. The torchtype ignition source is used for combustion in afterburner.

The air and fuel mixture is combusted again to increase the temperature of the fluid, which leads to high thrust after nozzle section as shown in Figure 1.21. This method increases the jet velocity and is known as *afterburning* and the device is called *afterburner*.

Pilots can start and stop the afterburner system in flight as per the requirement, and jet engines are referred to as operating wet when afterburner is used, and dry when no afterburner is used. The afterburner or augmenter is used for a few minutes mainly during take-off and sudden climb or other maneuver missions or to escape immediately.

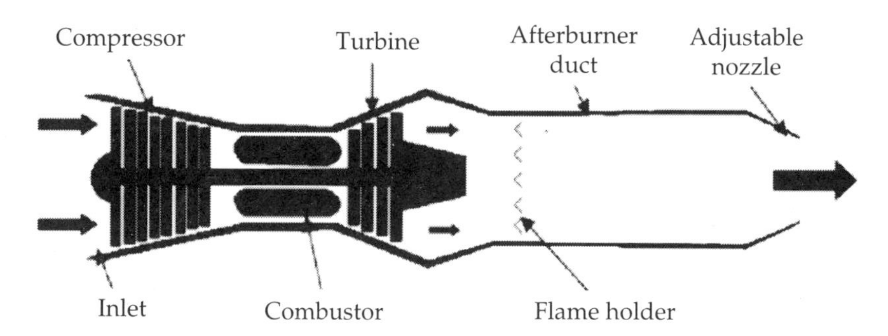

Figure 1.21 Afterburner.

Afterburners do produce a significant amount of thrust as well as a very large flame at the nozzle section of the engine. This exhaust flame may lead to shock diamonds, which is caused by shock waves formed due to the overexpansion of hot gas and also due to the small difference between ambient pressure and the exhaust pressure at the exit area of the nozzle. These imbalances cause oscillations in the exhaust jet diameter and the shock bands.

Low-bypass air gas turbine engines without actuating the afterburner system at supersonic speed leads to frozen state at nozzle exit. This is because of the lesser exit temperature than the thrust augmenter temperature and it may also lead to pressure inversion. In the combustion chamber, there are both the hot and cold losses, whereas in afterburner, there is only cold loss at all time, and hot losses appear only when it is actuated.

Water injection

It is another method of thrust augmentation. In this method, the air to be compressed in the compressor is enriched by injecting water or water-alcohol mixture at some location between inlet and exit of the compressor section in a very fine spray form or atomised form. By enriching the air, the load on the compressor becomes less for the required rise in pressure. The combustion chamber gets high dense air for combustion, which leads to increase in thrust.

Figure 1.22 explains the temperature and entropy cycle of turbojet afterburner engine. Process 0 to 5 is well-known from the previous studies which are general turbojet engine cycle. But, after the turbine process, the cycle is again extended. Process 5–6 is a constant pressure heat addition process, which takes place in thrust tube of afterburner. Then, isentropic expansion, i.e., process 6–7 takes place in the nozzle.

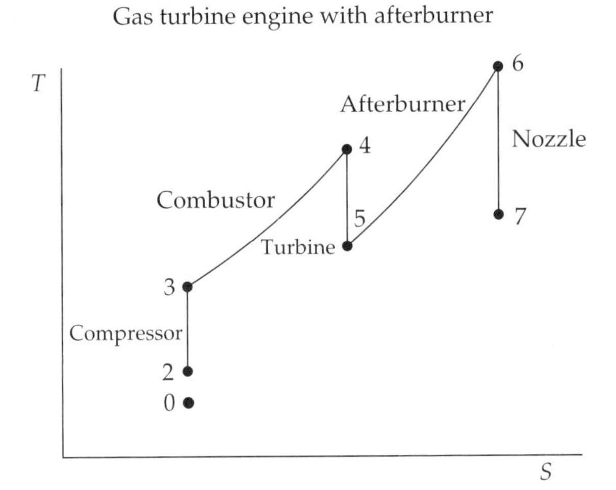

Figure 1.22 Afterburner *T-S* cycle.

1.18.1 Limitations of Thrust Augmentation

1. There are temperature constraints for the material used in afterburner and additional fuel has to be supplied.
2. Drastic reduction in temperature, increase in corrosion and the surging problem appear in water injection.

Afterburner is the most efficient method than any other method of thrust augmentation, as it has very less thrust-specific fuel consumption with less weight of subcomponents. Thrust vectoring can be easily inculcated in this method of thrust augmentation. *Thrust vectoring* means the thrust or jet at the exit of the engine can be directed in multiple directions.

1.19 DIFFUSION MASS TRANSFER

The *convection-diffusion equation* is a combination of diffusion and convection heat transfer equations. The physical quantities are transferred inside a physical system due to the following two processes:

1. Diffusion
2. Convection

The general equation is

$$\frac{\partial c}{\partial t} = \nabla.(D\nabla c) - \nabla.(\vec{v}c) + R$$

where, c is the variable, D is the diffusion coefficient, $\vec{v}$ is the average velocity of quantity which is moving, R is the source or sink, and ∇ is gradient.

The diffusion coefficient is a proportionality constant between molar flux due to molecular diffusion and the gradient in the concentration of the species.

Specific impulse

It is the amount of thrust generated by the combustion of a unit amount of fuel. Specific impulse, I_{sp}, can be calculated using the following equation:

$$I_{sp} = \frac{T}{\dot{m}_f} = \frac{\dot{m}(v-u)}{\dot{m}_f}$$

where, T is the thrust, $\dot{m}$ is the mass flow rate, v is the exit jet velocity, u is vehicle velocity and $\dot{m}_f$ is the mass flow rate of fuel.

Specific fuel consumption

It is the amount of fuel consumed to generate a unit amount of thrust force.

$$\text{SFC} = \frac{1}{I_{sp}}$$

Specific thrust

It is the amount of thrust generated by consuming a unit amount of air and fuel mixture. Specific thrust, T_{sp} is given by

$$T_{sp} = \frac{T}{(\dot{m}_a + \dot{m}_f)}$$

where $\dot{m}_a$ is the mass flow rate of air.

The performance parameters of a gas turbine engine are specific fuel consumption, specific thrust and specific impulse.

Maximum specific impulse can be found in pulse detonation engine followed by a gas turbine engine and then ramjet and scramjet engines. In gas turbine engine, again it can be classified as turbofan engine with maximum specific impulse followed by turboprop or turboshaft and then turbojet engine.

1.20 ACCESSORY SECTION

The engine is also linked with the accessory sections near compressor stages, which aid in driving the engine supporting systems and aircraft supporting systems like air conditioner motor, hydraulics pump, pneumatics pump, fuel pump, an electric generator, which acts as motor also, etc. The accessory section has an oil tank, drive shaft, inlet screen, generator and gearbox.

In gearbox, a set of bevel gears is used to drive an accessory shaft. The oil tank is for continuous lubrication of engine components; the driveshaft is an extension from the spool, which is used to drive the pump and generators. Generators are the electromechanical devices which produce electricity, and sometimes, act like motors in the starting condition and later act as generators. While starting an engine, the generator, which acts as motor, drives the spool taking energy from the ground utility or from another engine. Gearbox is used to have a control on shaft output from the engine RPM. This condition is more valid in turboprop engine, generator shaft work and turbo shaft engine.

1.21 AUXILIARY POWER UNIT (APU)

A large amount of electric power is required for lighting, entertainment and food preparation when aircraft is at ground (when main engines are not in ON mode). In addition to this, the main engine starting, cabin pressurising system and pneumatic air system should be taken care. Hence, auxiliary power unit is used. The auxiliary power unit is a small power plant or gas turbine engine located near the belly of aircraft as shown in Figure 1.23. An APU can provide electric power, air conditioning inside the cabin and start the main engines without the aid of any ground or portable power source.

Auxiliary power unit is a small size gas turbine engine. The engine length, number of compressor stages and fuel consumption are comparatively less than a normal gas turbine engine. The compressor stage used in the APU is centrifugal-type and the combustor is radial-type. Centrifugal-type compressor is used because it generates more amount of pressure than any other type of compressor in just one stage. It means only one set of compressor is enough to compress the fluid which also reduces the length of compressor section. But, the centrifugal compressor generates the pressurised fluid in a radial direction, which should be directed to the combustion chamber. Hence, radial-type combustion chamber is used.

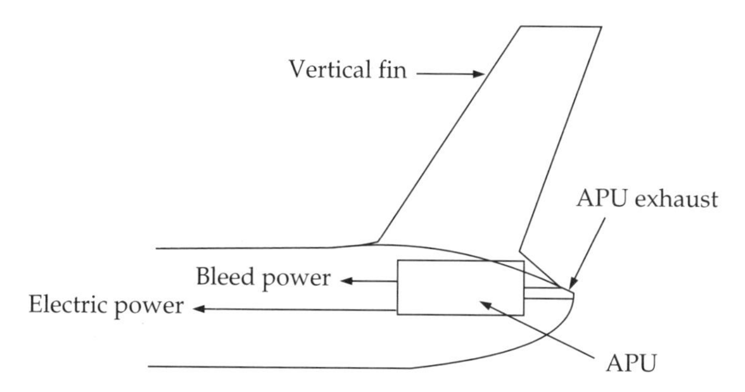

Figure 1.23 Auxiliary power unit.

The turbine extracts the maximum amount of energy from the hot fluid and remaining pressure energy is matched with the atmosphere using the nozzle. The APU is used to generate maximum shaft power and almost no exhaust jet power. The exhaust nozzle of the APU is designed only to match the exhaust pressure with the ambient pressure.

1.22 NOISE

Noise is unpleasant sound or disturbed sound. It is the standard noise abatement practice for an aircraft to overfly noise measuring stations at specified minimum altitudes. The unit used for noise measurement is effective perceived noise decibel (EPNdB).

There are mainly three types of noise producing sections in a gas turbine engine, namely, fan and compressor, turbine, exhaust. Maximum noise is at the exhaust and it is more at high turbulent region after nozzle due to shear action at the boundaries where high-speed turbulent jet and atmosphere meet. Low-frequency noise generates from high-frequency eddies of flow mixture and high-frequency noise generates from low-frequency eddies of flow mixture. The mixing of high-velocity hot gases from the engine with the low-velocity cold atmospheric air leads to the generation of low- and high-frequency noise. In turbofan engine at the nozzle section, mixing of hot gases from the hot section area and cold gases from the bypass section leads to a reduction in noise and it is in the range of 100–110 dB. This is because the cold flow of bypass section reduces the temperature and velocity of the hot gas coming out from the hot section.

Compressor and turbine noise come from the pressure field and turbulent wakes. The compressor and turbine blades create considerable wake region at varying mass flow rate, which leads to noise, and even if the turbo machinery load increases, then also the noise will come into the picture. At low thrust or taxing time, the noise can be heard. The noise will be even more if afterburners are used in turbojet or in low-bypass turbofan engines. Noise can be reduced by reducing the exhaust jet velocity or by increasing the mixing rate or by using the sound absorbing material or by reducing the region of mixing the exhaust with the atmosphere.

There are two type of noise suppressing nozzles, namely, lobe type and corrugated type. These nozzles break up the single exhaust stream into a number of smaller jets and increase the jet contact area. Both the types of nozzles reduce the size of eddies formed and they cannot be used for large diameter fan engine. Materials which suppress the noise convert acoustics energy into heat. The acoustics lining is composed of a perforated layer of thin steel, titanium or aluminum separated by a honeycomb of aluminum or stainless steel.

IMPORTANT QUESTIONS

1. Define aircraft propulsion. How is it different from aerospace propulsion?
2. What do you mean by thermodynamics? State the laws of thermodynamics.
3. Define Internal energy. Explain the modes of storing the internal energy.
4. What do you mean by entropy? What happens to the entropy of a system and the surrounding when heat is removed from the system?
5. Why is the specific heat value at constant volume smaller than the specific heat value at constant pressure?
6. Define thrust. Explain how it is generated.
7. What do you mean by efficiency of the engine? Explain the different types of efficiencies in aircraft engine.
8. Write the general form of steady flow energy equation. Also, write the steady flow energy equation for
   (a) Compressor
   (b) Combustion chamber
   (c) Nozzle

9. What do you mean by stagnation condition? Determine $\frac{T_0}{T}$ and $\frac{P_0}{P}$ at $M = 1$ (choked condition) and $\gamma = 1$.
10. What do you mean by compression ratio and efficiency in internal combustion engine?
11. Explain the different types of propellers.
12. What is the ignition order of single row 9-cylinder and two row 14-cylinder IC engine?
13. Explain the working procedure of turbo-charged internal combustion engine.
14. Name the cold section and hot section components of gas turbine engine.
15. Explain the factors affecting the thrust.
16. With a neat diagram, explain the specifications of the following:
    (a) Turbojet engine
    (b) Turboprop engine
    (c) Turbofan engine
17. Define thrust augmentation. Explain the types of thrust augmentation.
18. Explain the working of afterburner and plot a *P-V* and *T-S* diagram for a gas turbine with afterburner arrangement.
19. Define specific impulse, specific thrust and specific fuel consumption.
20. Define APU. Explain the functions of APU.
21. Write the noise producing components in gas turbine engine and radial engine.

CHAPTER 2

INLET, COMPRESSOR AND DIFFUSER

OVERVIEW

In this chapter, we will look into the leading components of a propulsion system like inlet, compressor and diffuser used in various propulsion systems in detail. These are known as pressurising components (static or dynamic in nature) and also known as cold components of the propulsion systems. The theory, classifications, working principle, design variables and working conditions of compressors used in gas turbine engines are also discussed to the understanding level.

The inlet, compressor and diffuses are the basic cold components of gas turbine engine. These components increase the pressure of air flowing through them. The efficiency of these components decides the thrust and overall efficiency of an engine. The working limit of these components depends on the input values of hot components which are discussed further.

2.1 INLET

The leading major part of any gas turbine engine (in turboprop engine after propeller) is inlet, which acts as diffuser. Hence, it is also called *diffusing duct*. The inlet is a duct where the pressure of air increases and velocity decreases along the inlet length. Inlet increases the pressure of air flow by virtue of vehicle's or engine's forward motion and increases its static pressure inside the intake manifold of an engine. The inlet is classified mainly into two types, namely, subsonic inlet and supersonic inlet.

The inlet duct transforms the kinetic energy of the air flow into pressure energy by internal compression or by external compression or by combination of both. For a supersonic flow, *internal compression* is one where the flow entering the inlet is through multiple shock waves or bow-like structures which increases the flow pressure and decreases the flow velocity, whereas the *external compression* means the supersonic flow is compressed just before it enters (outside the inlet) the inlet due to very high air velocity. These are the main compression methods seen in supersonic inlets.

In subsonic inlet, the subsonic flow which is flowing through the inlet duct expands, as its kinetic energy decreases and the pressure energy increases. This phenomenon is called *internal compression* or just *compression in subsonic flow*. The internal compression can be seen in both subsonic and supersonic types of inlets, whereas the external compression can be seen only in supersonic inlets.

Assuming no change in the cross-sectional area of inlet, the flow inside this parallel inlet duct is called *Fanno flow*. Fanno flow means the fluid flows in a constant area duct with friction in the absence of heat interaction.

The effect of wall friction on the fluid parameters is significant in long pipes and ducts. The wall friction makes the subsonic flow into sonic flow, where the flow velocity increases and its pressure, density and temperature decreases, and the supersonic flow becomes sonic flow where the flow velocity decreases and its pressure, temperature and density increases.

The inlet is a critical part which has a significant effect on engine efficiency and compressor efficiency. It means in the absence of inlet, the compressor is completely exposed to the upcoming flow from all the directions; the load on the compressor is more and hence, the power required to drive the compressor as more, which is nothing but less momentum thrust generation. Non-uniform flow from inlet may also cause disturbance to the compressor or compressor blade stall or surge, which can result in a reduction in compressor efficiency and severe mechanical damage due to blade vibration.

In inlet, there are losses in pressure due to inlet wall friction and shock wave effects on the flow at subsonic and supersonic speeds, but the resultant pressure is more than the ambient. Under static condition or very low forward speed that is during engine start up and taxing, the intake acts as a nozzle, where the air accelerates from zero or low velocity to the required velocity so that it reaches the compressor stage for compression with the required pressure. But, at high speeds, the inlet performs as a diffuser, thus decelerating the flow, and increases the pressure, which in turn reduces the compressor load.

Subsonic inlets consist of fixed geometry duct whose diameter progressively increases from entry to exit section like divergent duct, whereas in supersonic inlets, there is either fixed or variable geometry and their diameter progressively decreases like a converging duct, as shown in

Figure 2.1 (a), (b) and (c). The inlet is the only component of gas turbine engine based on which the mass flow of air entering the engine can be decided, or even we can say that based on the mass flow rate of air required, the inlet is designed and this leads to thrust calculations. Inlet provides sufficient air supply to the compressor with minimum pressure loss.

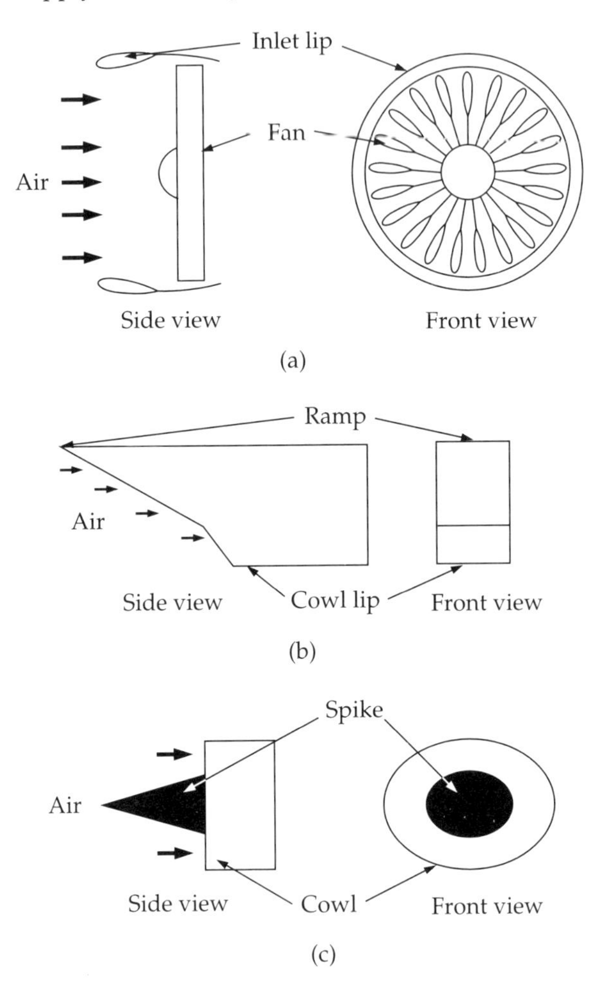

Figure 2.1 (a) Subsonic inlet, (b) Supersonic rectangular inlet and (c) Supersonic axisymmetric inlet.

The inlet design is very critical and it depends on few parameters.

2.1.2 Inlet Design Variables

1. Altitude envelope
2. Engine location

3. Inlet total pressure ratio
4. Integration of fan, propeller, and different flow paths
5. Flow field interaction between nacelle, pylon and wing
6. Diffuser wall diverging angle.

The inlet is designed based on the requirement of air mass flow rate. The mass flow rate of air is maximum at low altitude and high speed. If the mass flow is less available, it means thrust generation is less, which is generally at cruise condition at higher altitude. Hence, the thrust value and operating altitude decide the mass flow rate of air with respect to the entry area. Inlet lips are designed to have increasing capture area, which increases the mass flow.

The engine location also affects the inlet design, as the thrust required is directly proportional to the mass flow of air which is getting inside the engine. The air should have no disturbance from any other aircraft component before getting inside the engine.

The wing mount engines can have different inlet sizes, but the fuselage mount engine has got a restriction in the inlet size from structural and aerodynamic point-of-view. If the engine is mounted in fuselage where the inlet is very close to the nose of an aircraft, then the disturbance due to flow separation from the nose or front part of the fuselage will affect. If the engine is wing mounted, then the flow separation will be very less and inlet performance will be very good along with some drawback from aerodynamic point-of-view like less maneuvering freedom. The wing mounted engines also get some amount of flow disturbance because of nacelle and pylon components.

The inlet design also depends on the requirements of the compressor, i.e., the inlet total pressure ratio or the reduction in the flow velocity. The inlet duct length and inlet duct area ratio depend on the increase in the flow pressure or decrease in the flow velocity as per the compressor input requirements. The excess increment in the inlet area ratio also leads to flow separation followed by a wake in the inlet, which means reduction in efficiency.

In a turbofan engine, the inlet design varies, as there is not enough length for inlet duct because of the fan, and also, the flow properties requirement for compressor are different. Similarly, for turboprop engine, the propeller has an impact on inlet flow. In some supersonic fighter aircraft, only one engine is used, but two inlets are provided for that. These inlets should be designed to provide the required acceptable value of flow to the engine.

2.2 COMPRESSOR

The second leading major component in the gas turbine engine is compressor, which is used to increase the pressure and reduce the velocity of the air flow which is taken into the engine. A device which makes the water to flow from its lower level to higher level is generally called hydraulic pump or water pump, but if it pumps gaseous fluid like air, then depending on the pressure rise, it can be classified mainly into three types. Up to 0.07 bar relative pressure rise at downstream, it is simply called *fan*, between 0.07 bar–0.3 bar relative pressure rise, it is called *blower* and above 0.3 bar relative pressure, it is called *compressor*.

The key factor that affects the compressor efficiency and engine efficiency is the compressor pressure ratio. *Compressor pressure ratio* is the ratio of the air pressure at compressor discharge to the air pressure at the compressor inlet. Higher the pressure ratio means higher the efficiency of the compressor as well as that is of engine. The compressor pressure ratio depends on the compressor loading parameter, the rise in temperature in the compressor stage and the absolute velocity of the fluid across the compressor rotor blades. The compressor blades also undergo thermal expansion, which affects the blade clearance ratio; hence, compressor disk and drum are cooled or heated to maintain the blade clearance ratio.

2.2.1 Types of Compressor

The compressors are classified into two main types, namely, centrifugal compressor and axial flow compressor.

Centrifugal Compressor

A *centrifugal compressor* is one where the air enters the compressor in the axial direction and leaves it in the radial direction. The centrifugal action of the blades of compressor on the fluid makes the fluid pressurised.

What would happen if we design a compressor where the air enters radially and comes out axially? The designing of the inlet for the compressor is very difficult and we cannot use centrifugal principle or centrifugal action for compressing the fluid. It just cannot be compressed.

Centrifugal compressor with single or double entry impeller was used in early days. In this, the airflow gets inside the compressor from the compressor hub or compressor eye in the axial direction. The hub or eye is a nozzle or convergent shaped duct, which helps in increasing the flow velocity. The accelerated airflow is made to flow across the compressor blade or impeller rotating at high revolution. The impeller blade diameter increases along the compressor length, as shown in Figure 2.2. The output flow from the compressor blades or impeller is with high kinetic energy.

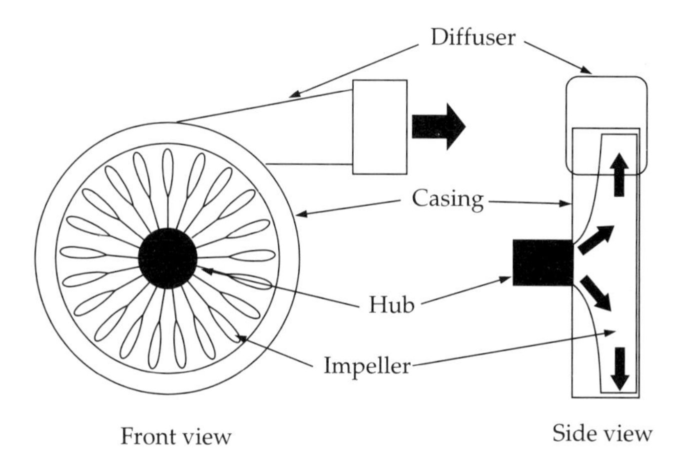

Figure 2.2 Centrifugal compressor.

In the hub, only kinetic energy of the air increases, and in the impeller, kinetic energy and static pressure of air increase. When the compressor starts from its static position, the impeller enhances the kinetic energy only, with no increase in static pressure, but at one stage, when the downstream pressure builds up in the diffuser, then that pressure makes centripetal force on the fluid, which leads to increase in static pressure of fluid. Hence, impeller blades are known for kinetic energy and static pressure enhancer. The airflow is made to flow through the diffuser section where the air velocity is reduced, transforming the kinetic energy into pressure energy.

A centrifugal compressor is limited to approximately 5:1 to 6:1 pressure ratio. Now, up to 10:1 pressure ratio can be achieved in a single stage centrifugal compressor. This type of compressor is largely used in small size gas turbine engines because of its considerable features like small length, single stage, low cost, a wide range of mass flow rate, easy to manufacture, but it has large frontal area and low efficiency.

The maximum mass flow rate through the impeller eye is approximately 35 kg/s. Due to the impeller blade action, angular motion of the air increases static pressure and acceleration. Shroud, which is part of the impeller, is used between impeller blade and casing to avoid flow over blade or reverse flow, which leads to windage loss. For a general condition, the mass flow rate and the compressor RPM may go up to 35000 RPM.

In a centrifugal compressor, the efficiency drops at a rapid rate because of excessive speed of the impeller, which leads to shock formation near the impeller tip. The inducer is one more part of the impeller blade, which enhances the angular velocity of the fluid.

In centrifugal compressor, the change in angular velocity is almost zero for a given RPM. Hence, the pressure rise in the compressor is only due to the initial velocity and radius of the impeller. As the change in the radius of the compressor is positive, then the centrifugal compressor pressurises the fluid positively, and as the change in radius of the turbomachine is negative, it extracts the energy, hence called *radial turbine*.

In a centrifugal compressor, the pressure rise is due to the angular momentum and absolute velocity component. For centrifugal compressor, change is absolute velocity component is zero and for axial flow compressor, the change in the radius is zero.

Axial Flow Compressor

In axial flow compressor, the air enters the compressor axially and leaves axially. With this type of compressor, multiple stage and multiple set arrangement can be made.

Each stage of the compressor has one rotating member and one non-rotating or stationary member, as shown in Figure 2.3. The rotating part is called *rotor* and non-rotating or stationary part is called *stator*. In the axial compressor, the impellers are not present, whose diameter varies from hub to tip, and also, there is no centrifugal action by the compressor blade to increase the velocity or pressure.

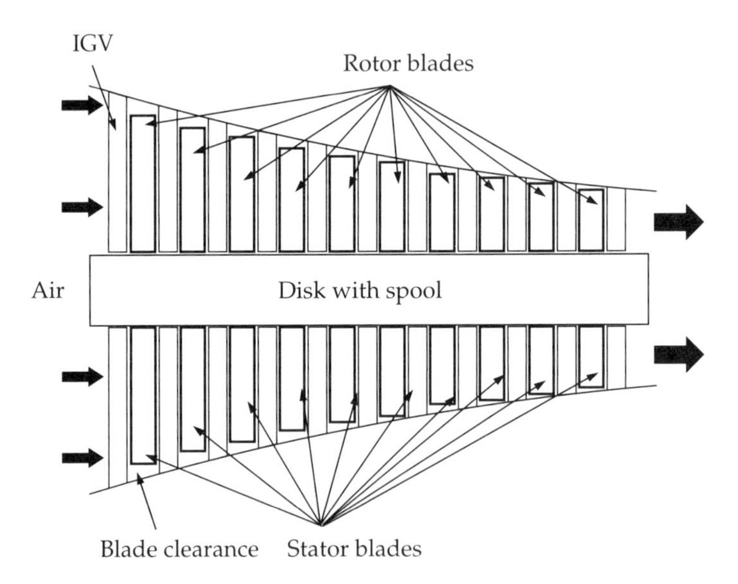

Figure 2.3 Axial flow compressor components.

In axial compressor stage, the inlet guide vane (IGV) guides the flow initially to enter the rotor blade at the required angle with minimum disturbance. The rotating rotor component increases the velocity of the fluid. The high-velocity fluid is made to go through the stator stage where the kinetic energy of the fluid is transformed into pressure energy. If we consider the cross section of a stator, any two vanes of stator together make a divergent shape, hence, the flow pressure increases in the stator.

Each compressor stage is having a rotor as leading component followed by stationary component. Each vane (stator part) and blade (rotor part) is of airfoil shape, which helps the flow to flow across the compressor with less boundary layer separation. To achieve more pressure ratio, more number of compressor stages can be used which rotate at the same speed. This combination of many compressor stages together is called *compressor set*. Multiple compressor sets can also be used which rotate at different speeds to achieve more pressure ratio using less compressor stages and occupying less length or space.

In multiple set axial flow compressors, the total pressure ratio is not the sum of pressure rise in each stage, but it is the product of pressure rise in each succeeding stage. To maintain the required velocity with pressure increase, the diameter of the compressor set is gradually decreased with each succeeding compressor stages from low- to high-pressure end, as shown in Figure 2.4. To have efficient combustion, the discharge velocity is maintained close to the inlet velocity of the compressor to avoid excessive diffusion.

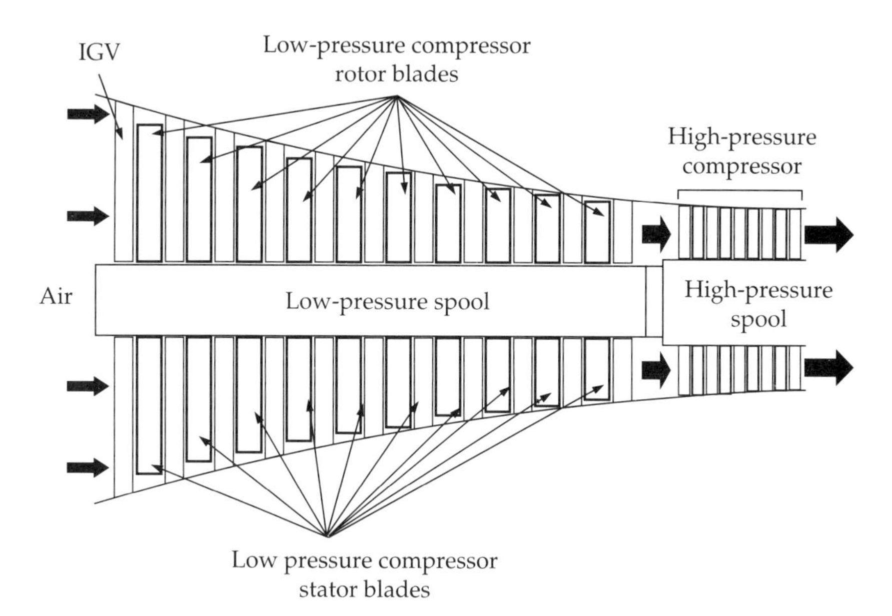

Figure 2.4 Multi-stage axial compressor.

A compressor can also be classified based on its RPM and pressure rise. In two-spool or triple-spool gas turbine engines, we can see two types of compressor stages. One achieves low pressure, whereas another achieves high pressure and these are called *low-pressure compressor* and *high-pressure compressor*, respectively. Low-pressure compressor runs at low RPM and high-pressure compressor runs at high RPM. In the high-pressure compressor, the distance between the rotor blades or stator blades is relatively less, and also, the distance between the blade tip and the casing is relatively less.

The rotor blades are attached to a disc (hub) mounted on a central shaft or spool forming a compressor rotor, and static vanes are fastened to the inner surface of compressor case. As the rotor turns, the air is pushed across rotor blades. Each stage increases the air pressure until it leaves the compressor at significantly high pressure. Air becomes denser and acquires less volume as pressure increases. To prevent the drastic velocity reduction, the cross-sectional area is reduced.

Stator vane also helps in correcting the deflected air flow and conveys it at a required angle to the next stage. The last row of stator makes the airflow straight and removes the swirl from the air prior to the combustion chamber. To avoid the discontinuous flow and blade stall phenomenon, each stage is designed to increase small amount of pressure.

Every rotor needs to have a different design as per its flow requirement, if there is no stator for every rotor. The first set of compressor has an inlet guide vane prior to rotor, which guides the flow to enter the rotor at right angle. Compressor guide vanes are thin and hollow as compared to compressor rotor blades. If a compressor set has more than 8 stages then after every 5th stage, inlet guide vane is used to guide the flow to the next rotor.

If compressor RPM is constant and mass flow rate of air decreases with a change in altitude, then the angle of attack of stator vanes can be varied using actuators, which maintains the compressor efficiency constant and it is true in opposite condition also. If the vanes angle of attack is increased beyond the critical point, it will lead to a stall condition. To increase the stall margin when operating under a condition other than compressor design speed, the inlet guide vanes and multiple stage stator vanes are made variable using some actuating system, which is directly linked with the fuel control lever.

The amount of fuel that can be burnt is proportional to the mass flow rate of the incoming air and this incoming mass flow rate of air and its total temperature are a function of compressor design. In a single stage axial compressor, the pressure rise is up to 1.3 bar to 1.4 bar at 88%–90% efficiency. The maximum mass flow rate that an axial compressor can compress is around 700 kg/s.

The rotor blades increase the kinetic energy of the fluid, which is transformed by the stator vanes to increase its static pressure. Seals are provided near the tip of the blade to restrict the reverse flow between the stage and set of compressors. Based on the total temperature rise in compressor, the number of compressor set is determined.

The pressure rise to be produced inside the compressor is decided based on the temperature rise inside the compressor. In axial flow compressor, the windage loss is very less as compared to centrifugal compressor.

Table 2.1 shows the differences between axial flow compressor and centrifugal compressor.

TABLE 2.1 Differences between Axial Flow Compressor and Centrifugal Compressor

S.No.	*Axial flow compressor*	*Centrifugal compressor*
1.	Exit flow direction is parallel to the axis of rotation.	Exit flow direction is radial to the axis of rotation.
2.	Maximum pressure ratio at each stage is 1.3 to 1.6 times.	Maximum pressure ratio at each stage is 5 to 6 times.
3.	Isentropic efficiency is 88%–90% for a mass flow of more than 6 kg/s.	Isentropic efficiency is 80%–82% for a mass flow of more than 6 kg/s.
4.	Flexibility of operation is limited.	Flexibility of operation is more.
5.	Starting torque is high.	Starting torque is low.
6.	It is suitable for multistage.	It is not suitable for multistage.
7.	There is small cross-sectional area.	There is large cross-sectional area.

2.2.2 Efficiency of Compressor

The efficiency of the compressor, η_C is given by the ratio of actual rise in temperature in the compressor to the ideal temperature rise in the compressor. The compressor efficiency can be expressed as shown in Eqs. (2.1) to (2.3).

$$\eta_C = \frac{\dot{m} C_p (T'_{02} - T_{01})}{\dot{m} C_p (T_{02} - T_{01})} \tag{2.1}$$

$$\eta_C = \frac{T'_{02} - T_{01}}{T_{02} - T_{01}} \tag{2.2}$$

$$T_{02} - T_{01} = \frac{T_{01}}{\eta_C}\left(\frac{P_{02}}{P_{01}}\right)^{\frac{\gamma-1}{\gamma}} \tag{2.3}$$

where T_{01} is the stagnation temperature of the flow at the compressor inlet. T_{02} is the stagnation temperature of the flow at the compressor exit (ideal case) and T'_{02} is the stagnation temperature of the flow at the compressor exit (Actual case). P_{02} is the fluid stagnation pressure at compressor exit and P_{01} is the fluid stagnation pressure at compressor inlet.

2.2.3 Compressor Design Parameters

1. Number of spools
2. Spool RPM
3. Number of stages and spacing between them
4. Number of blades on each stage in axial flow compressor
5. Pressure ratio per stage
6. Air mass flow rate
7. Compressor efficiency
8. Flow path shape
9. Surge margin

In the compressor, the pressure can be increased in multiple steps. If the pressure requirement is very less, then only one type of compressor set can be used, which is driven using single spool connected to turbine. But, if the pressure rise requirement is more, it leads to two or three different types of compressor sets, which rotating at different RPM, mounted on separate spools generally called high-pressure compressor, intermediate-pressure compressor and low-pressure compressor. This technique helps in reducing the number of compressor stages or the length of compressor area.

Compressor design depends on the pressure rise required at the exit and the number of blades in each rotor member. The amount of airflow to be compressed depends on the compressor diameter, rotor RPM, and also on the number of blades. If the RPM of the rotor is less, the kinetic energy generation in the rotor will be less, whereas if the number of blades on the rotor hub is less, then the load on the each rotor blade will be more. The rotor blades are fixed to the rotor hub at a particular angle considering the stall and surge margin.

The compressor design takes care of the efficiency of the compressor also by achieving the ideal temperature rise during compression. The combination of axial compressor followed by the centrifugal compressor is used to avoid the individual limitations.

2.2.4 Bleed Air Utilisation

Bleed air is the compressed air which is taken out after the compressor section at multiple sections, depending on the required fluid pressure and temperature, as shown in Figure 2.5. It is used to cool the engine's critical hot sections, in sump pressurisation or purging of engine cavities and pneumatic systems like pressurisation, anti-icing of cowl and wing and air conditioning.

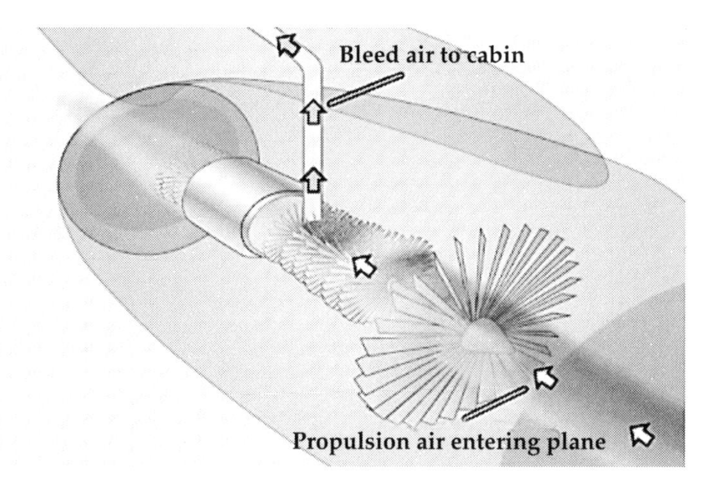

Figure 2.5 Bleed air taking out after compressor section.

The air extraction varies as per the demand. The loss in thrust is directly proportional to bleed air extraction at various sections. Bleed air is taken out from the bleed port of diffuser after compressor stage for de-icing, anti-icing and pneumatic engine start.

2.2.5 Compressor Stall

Like an aircraft wing, compressor blades are also in airfoil shape and have an acute angle of attack. Compressor stall happens when the angle of attack of rotor blade goes beyond critical angle of attack. It leads to slowing down of air supply followed by reverse flow. It can also happen because of turbulent air at the inlet or sudden acceleration of aircraft.

When air reaches unstable condition, stall results. This phenomenon is found more in high-pressure axial flow compressor. When the relationship between pressure, velocity and compressor ratio is altered, the angle of attack of the blade is changed by changing the flow angle using variable guide vanes. If relation becomes incompatible, the angle of attack becomes excessive, resulting in stall and flow becomes very turbulent. During ground operation, at low thrust, mild stall can be occurred resulting in rumble and chugging.

If the temperature of the engine gas increases or RPM fluctuates drastically or engine pressure ratio decreases, stall will encounter. If the loads acting on the compressor stage are high and lower than the critical value, then also it leads to stall. The highly loaded stages are made stall-free by opening the bleed valve at the stage section. If the loads acting on the compressor are beyond the critical value, then it leads to surge. Variable bypass valves are also used to avoid stall and compressor inlet temperature.

2.2.6 Dual Compressor Theory

Dual compressor theory provides greater operating flexibility over a wider speed range and at higher compression ratio. It has improved stall margin. It also called *twin-spool compressor*, i.e., high-pressure compressor and low-pressure compressor together in an axial type of compressor. The energy consumed to drive high-pressure compressor is more as compared to energy consumed to drive the low-pressure compressor. Dual compressors are used where large airflow with high pressure is in demand. Sometimes, triple-spools are also used, i.e., high-pressure compressor set, intermittent or intermediate pressure compressor set and low-pressure compressor set. When we look at the dual centrifugal compressor stage, two centrifugal compressors mounted on the same shaft have same angular velocity, but the inlet pressure is different. The blade shape and radius might be different, which leads to the different amount of pressure rise. The dual compressor theory for centrifugal compressor does not hold good for gas turbine engine because the radial output from one compressor should be made axial to another. This process leads to increase in the diffuser with diaphragm length and low efficiency. The centrifugal compressor can be used with axial type compressor as multistage by arranging the centrifugal type compressor after axial.

2.2.7 Variable Inlet Vane and Stator

The variable inlet guide vane technology is incorporated in the year 1954. The vanes of stators are hydraulically actuated and connected in parallel to the fuel control system. At low engine speed, these are at minimum angle position or the diffusing angle of the stator blades is more. As engine accelerates, they adjust the stator vane angle towards maintaining the exit pressure from the compressor. This flexibility is given at the compressor stage so that the compressor adjusts itself based on the requirements and environment conditions. The inlet induced surge on the compressor blade can be overcome by this flexibility provided to the compressor stator. For example, in the J-79 engine, the stator blades are attached to the actuator ring to get the adjustable motion.

Compressor loading parameter, C_{LP}, is the amount of load taken by each compressor stage with its highest efficiency. It means this parameter is used to find the number of compressor stages and sets required with a relative increase in temperature, i.e., Eq. (2.4), where C_P is specific heat of air at constant pressure, ΔT_s is the change in stage temperature and U is the peripheral velocity or blade velocity.

$$C_{LP} = \frac{C_P \Delta T_S}{\frac{1}{2} U^2} \tag{2.4}$$

Generally, the compressor loading value is fixed and with different absolute velocity component of fluid flow, the rise in temperature value is found using Eq. (2.4). Based on the rise in temperature, the number of compressors is fixed.

2.3 DIFFUSER

The *diffuser* is an internal flow structure or duct in the aero engine, which is used to decrease the velocity by increasing the pressure of the fluid flowing through it. This device may be used before compressor stage (inlet) and after compressor stage also. The exit air velocity from the compressor may reach somewhere around 500 ft/s, which is too fast for combustion. Hence, the diffuser is used between the combustion chamber and compressor stage. The diffuser is also used to take out the compressed air as bleed after the compressor, as the ports are available easily on the diffuser circumference section. Figure 2.6 shows the typical structure of a diffuser used in gas turbine engine which has bleed ports on its periphery.

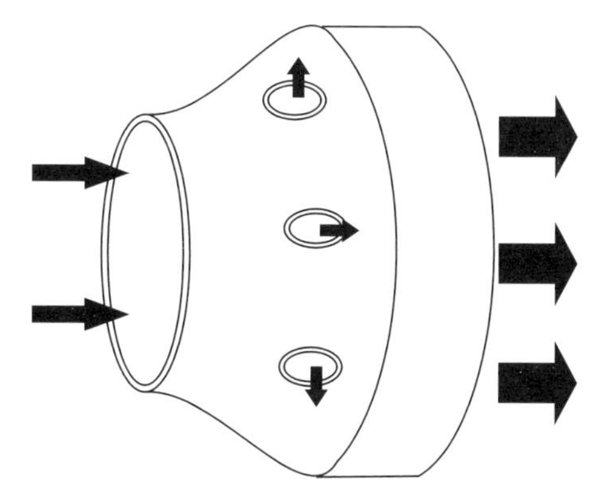

Figure 2.6 Diffuser.

IMPORTANT QUESTIONS

1. What do you mean by inlet? Write the functions of inlet.
2. What is the difference between subsonic inlet and supersonic inlet?
3. Explain internal compression and external compression in supersonic inlet.
4. Define Fanno flow.
5. Explain the inlet design variables.
6. Define compressor pressure ratio. How does compressor pressure ratio affect the engine efficiency and compressor efficiency?
7. Explain the working of centrifugal compressor.
8. With a neat diagram, explain the centrifugal compressor components.
9. Is multi-staging of centrifugal compressor possible? Justify your answer.
10. What are the different parts of axial flow compressor? Explain each of them.
11. Explain the working of low-pressure compressor and high-pressure compressor.
12. Define and explain compressor stall and compressor surge.
13. Write the difference between axial flow compressor and centrifugal compressor.
14. Explain the compressor design parameters in detail.
15. What do you mean by bleed air? How does bleed air system work?
16. How do you arrange the following components to have the best efficiency out of the arrangement? Low-pressure axial flow compressor, high-pressure axial flow compressor and centrifugal type compressor.
17. What do you mean by inlet guide vane? Explain how variable inlet guide vane works.
18. Define diffuser. Explain how diffuser helps the compressor.
19. Explain the working of diffuser with bleed ports on its circumference.

CHAPTER 3

COMBUSTION CHAMBER AND NOZZLE

OVERVIEW

In this chapter, we will look into the hottest part of all the types of propulsion system, i.e., combustion chamber. We will look into the design, operation and performance aspects of combustion chamber used in mainly gas turbine engines. Classifications and instabilities of the combustion chamber are also discussed in this chapter. In this chapter, after combustion chamber, we will discuss nozzle, its classification and its operation in detail.

Combustion chamber is the hottest part of every engine. The chemical energy of fuel is converted into heat energy, which in turn increases the enthalpy of the working fluid. Nozzle is the thrust generating component of engine. Nozzle can have cold flow and hot flow across it. The high-pressure fluid inside the nozzle creates thrust by forcing the surface of the nozzle. The details of these processes are discussed further.

3.1 COMBUSTION CHAMBER

Combustion is a thermo-chemical process where the chemical energy of the fuel is converted into heat energy. In Brayton cycle, process 2–3 is combustion process which is at constant pressure and in Humphrey cycle, process 2–3 is combustion process which is at constant volume.

In gas turbine, the combustion process involves the following:

1. Formation of reactive mixture
2. Ignition
3. Flame propagation
4. Cooling of combustion chamber

Generally, combustion chambers in typical gas turbine engines are designed, which get an inlet air velocity not exceeding 80 m/s at design load. The mixture of air and fuel or oxidant and fuel is made as per the stoichiometric principle to initiate the combustion with a calculated ignition source.

The fuel injector must inject the fuel as very fine droplets or in mist form in the combustion chamber, where the primary air entering the combustion chamber mixes with it, rather than distracting it from the ignition source point. The fluid flow in the combustor is known as *Rayleigh flow*. It is the flow in a constant area duct with heat transfer without friction. The heat can be added to or removed from the duct.

The energy required for the ignition source is measured in terms of joules, and selection of the location of ignition source or the distance between the premixed fuel-air and ignition source inside the combustion chamber is very crucial. Based on the fuel injecting direction, the angle of fuel injection cone and the air supplied (which creates turbulence), the ignition source location is decided.

A typical ignition source for a gas turbine engine is the high-energy capacitor discharge system, which provides 60~100 sparks per minute with several inches spark length. The draining system accomplishes the draining of unburnt fuel after shutting down. The combustor consists of an outer casing with a perforated inner liner, which protects the casing and cowling from the heat produced in the combustion chamber.

The compressed air from compressor stage is sub-classified in the combustion zone as combustion initiating air (primary air), combustion propagating air (secondary air) and the combustion flame cooling air (tertiary air supply). The combustion chamber has airports or passages at its circumferential area.

The combustion propagating air enters the combustion chamber through the air passages and helps the combustion flame in propagating with better turbulence. This flame propagating air leads to flame turbulence inside the combustion chamber and improves the combustion process by having complete combustion of fuel. The increase in turbulence level leads to the considerable amount of pressure loss inside the combustion chamber, and the decrease in turbulence level leads to incomplete combustion and also increase in combustion chamber length. The combustion propagating air and tertiary air help in cooling of the combustion chamber by creating a thin film of air around the combustion chamber. Tertiary air mixes with the combustion products at the end of the combustion chamber and reduces the combustion exhaust temperature, which usually affects the turbine blades.

The fuel-air mixture is classified as a rich mixture, a lean mixture and a stoichiometric mixture. In a thermo-chemical process, for complete combustion, we need some amount of fuel and some amount of air. The amount of air required to have a complete combustion of a quantity of fuel is known as *stoichiometric ratio*. The ratio of stoichiometric air-fuel ratio to the actual air-fuel ratio is known as *equivalence ratio*. If the amount of air is less than the amount of fuel towards the combustion equivalence or in a complete combustion process, if there is excess amount of fuel, then it is called *rich mixture,* which leads to incomplete combustion (equivalence ratio more than 1). If the amount of fuel is less than amount of air towards the combustion equivalence or in a complete combustion process, if there is excess amount of air, then it is called *lean mixture,* which leads to complete combustion with less combustion temperature (equivalence ratio less than 1). If the amount of fuel is same as amount of air towards combustion equivalence, then it is called *Stoichiometric mixture,* which leads to complete combustion with high combustion temperature (equivalence ratio equal to 1). *Equivalence ratio* is the measure of richness or leanness of the air-fuel mixture.

When the chemical reaction is balanced, then the equivalence ratio is unity. We know that by substituting the molecular weights, in the chemical equation, the mole fractions get converted into mass fractions. At equivalence ratio 1 or at the stoichiometric ratio, the mass fraction of oxidiser to fuel is in the range of 15–60 depending on the type of hydrocarbon fuel and oxidiser. If oxygen is used as oxidizer, then the fuel-air ratio is close to 15 and if air is used, then the oxidiser to fuel ratio is close to 60.

The combustion reaction of liquid fuel generally takes place at gaseous and liquid phase (two phase). In the premixed condition where the liquid fuel is mixed with the gaseous oxidiser and injected through the injector as a fine spray, it has small liquid fuel droplets mixed with gaseous state fuel. Such a mixture is called *two-phase mixture.* In a two-phase mixture, if the droplets size is more, it leads to incomplete combustion of the fuel and deposition of fuel on the combustor wall; hence, the droplet size should be very fine. The two-phase problem can be resolved by increasing the downstream pressure of fuel or by reducing the exit area of the injector.

Flame speed is an important factor inside the combustion chamber, which depends on the injection speed and area available for combustion. The fuel droplet collision should be avoided after the injection process which gives closer value to the complete combustion heat release.

The compressed air from compressor should mix with fuel in turbulent fashion which leads to fuel diffusion. This phenomena leads to complete fuel atomization. Turbulent vortices get created around the flame which helps the flame from acoustics and flame extinguishing instabilities. In combustor, 20–25% of incoming air is primary, which passes through the swirl vanes, which gives the radial motion at 5–6 ft/s velocity, whereas 75–80% of the secondary incoming air is at several feet per second.

The amount of fuel to be added in combustion chamber depends on the temperature rise required across the combustor. The secondary air is directed radially into the combustion chamber. So, the flame stabilisation can be enhanced by creating toroidal vortex recirculation.

What would happen if we place a combustor before compressor? In the discussion of gas turbine engine, combustion chamber cannot be placed before compressor because it is difficult to pressurise the hot combusted gas, as there is a limitation to compressor and requirement of energy for compressor is more. The compressor material selection must be advanced, and also, the combustion chamber efficiency will come down if the input charge is not pressurised.

3.2 FACTORS AFFECTING THE COMBUSTION CHAMBER DESIGN

Following factors affect the design of the combustion chamber:

1. The temperature of the combustion chamber products must be comparatively low.
2. The exit temperature must be known or within calculated range to keep turbine and its blade safe from thermal effect.
3. Maximum air velocity must be 30–60 m/s at the entry of combustion chamber.

3.3 FACTORS AFFECTING THE COMBUSTION PROCESS AND PERFORMANCE

The following factors affect the combustion process and its performance:

1. Pressure loss: We know that turbulence is necessary for rapid and complete combustion, but turbulence depends on the compressed air impact inside the combustion chamber. The turbulence causes the pressure loss in the combustion chamber. Pressure loss is caused by two factors, namely, pressure loss due to the friction of turbulent flow and the acceleration accompanying heat addition. The combined pressure loss is due to both the heat addition and friction, and the sum of pressure loss is determined separately as cold losses and hot losses. In the combustion chamber, the loss in stagnation pressure is due to heat addition or the momentum produced by the exothermic reaction mainly and also, due to aerodynamics resistance of flame stabilisation and mixing devices. Pressure loss can be minimised by using a large combustion chamber with consequent lesser velocities, which is feasible in the industrial unit, where the size is not critical. Equations (3.1) and (3.2) represent the expression for pressure loss factor (PLF).

$$\text{PLF} = \frac{\Delta P_0}{m^2/2\rho_1 A_m^2} \tag{3.1}$$

where, ΔP_0 is drop in stagnation pressure, m is the mass flow rate of fuel, ρ_1 is the density and A_m is the combustion chamber area.

$$\text{PLF} = K_1 + K_2\left(\frac{T_{02}}{T_{01}} - 1\right) \tag{3.2}$$

The pressure loss factor is the sum of cold loss and hot loss as shown in the above equations.

2. Combustion intensity: The size of the combustion chamber is determined by the requirement of the rate of heat release. The nominal heat release rate, Q, is expressed in Eq. (3.3).

$$Q = \dot{m}f\Delta H_c \tag{3.3}$$

where $\dot{m}$ is the mass flow rate of air, f is the fuel to air ratio and H_c is the lower calorific value of fuel.

$$\text{Combustion intensity} = Q/(\text{Combustion volume} \times \text{Pressure}) \tag{3.4}$$

The quantity 'combustion intensity' has been introduced to take an account of foregoing effects. In aircraft, the combustion intensity is in the range of 75×10^5 kJ/m^{3}hr to 150×10^6 kJ/m^{3}hr, where the industrial engines may be ten times of this value, and for rockets, it is in the range of 260×10^5 kJ/m^{3}hr to 260×10^6 kJ/m^{3}hr. The increase in the combustion intensity means an increase in the heat release rate, which leads to increase in the efficiency. The heat release rate varies from fuel to fuel and it is maximum for hydrogen fuel.

3. Combustion efficiency: Combustion efficiency is the ratio of the theoretical fuel-air ratio for a given increase in temperature to the actual fuel-air ratio for the same increase in temperature. The maximum combustion efficiency in gas turbine engine is 99% and the lower value of pressure loss is 2%–8%. The combustion efficiency, CE, for a given rise in temperature ΔT is expressed in Eq. (3.5).

$$\text{CE} = \eta_c = \left\{\frac{\text{Theoretical fuel-air ratio}}{\text{Actual fuel-air ratio}}\right\} \tag{3.5}$$

3.4 REQUIREMENTS OF COMBUSTION CHAMBER

1. Complete combustion of fuel must be achieved.
2. Total pressure loss must be very small.
3. Carbon deposition or soot formation should not happen in the downstream components.
4. Ignition must be reliable.

5. Controlled temperature and velocities must be there at the turbine inlet.
6. Volume and weight must be within the limits.

3.5 COMBUSTION PROCESS

1. Fuel injection
2. Proper mixing of fuel droplets and air
3. Atomisation, evaporation of fuel droplets
4. Ignition
5. Breaking the fuel molecules into lighter fractions
6. Mixing of combusted gases with the secondary and tertiary air supply
7. Having a controlled chemical reaction between fuel and air, which leads to the generation of heat and byproducts
8. 15%–20% primary air entering the combustion chamber for rapid combustion
9. Introduction of 40% of air secondary air through the ports
10. Mixing the remaining amount of air with the products of combustion to cool the products before turbine

With the change in altitude, the density and mass flow of air decrease compared to sea level condition, but the air-fuel ratio is maintained constant to have less variation in turbine entry temperature throughout the flight envelope. Gas turbine engines may also be designed for dual fuel operations with normal operation on natural gas and an option to switch to liquid fuels for a short period.

3.6 TYPES OF COMBUSTION CHAMBER

There are three types of combustion chamber, namely, can type, annular type and can-annular type or cannular type. These are described in the subsequent sections.

3.6.1 Can Type Combustion Chamber

In general, for understanding, we can say that individual cans are used to carry the combustion. Each can is provided with the fuel injector and ignition source separately. The high-temperature high-pressure air from the compressor is taken into the combustion chamber and split into the primary air (20%) and secondary air (80%) using elbow flange joint. The primary air is used for initiating the combustion process at primary zone, whereas the secondary air is used at intermediate levels after primary zone for diluting the combusted gas so that the combustion output to the turbine has a reduced temperature ranging from 1100°C to 1200°C and also, for cooling the combustion chamber internally to avoid thermal effects. The combustion chamber cans are not connected to each other internally, as shown in Figure 3.1. The volume or space available for combustion is comparatively less, hence, the loss due to wall friction and wall conduction will be more. It also leads to rapid increase in combustion casing or flame tube temperature.

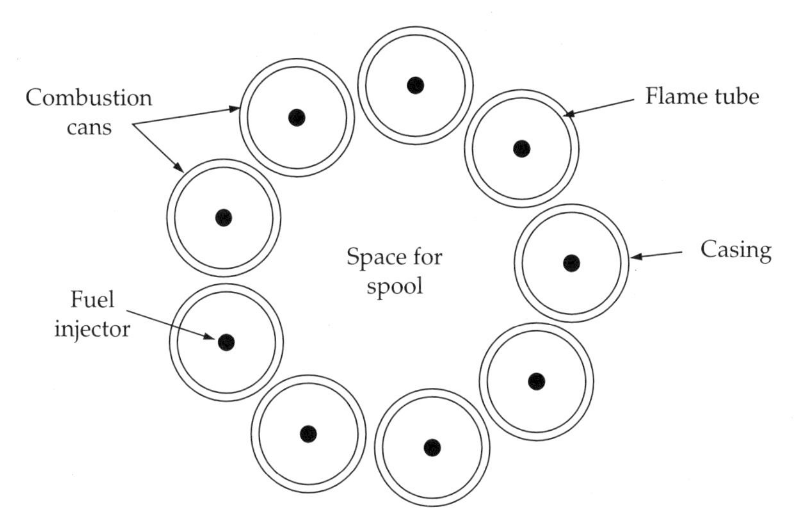

Figure 3.1 Can type combustion chamber.

3.6.2 Annular Type Combustion Chamber

Annular type combustion chamber is used when the approaching velocity of air is comparatively very high and heat is supplied to the same without reducing its velocity. This combustor has annular flow of air and fuel is injected inside the combustion chamber with very less obstacle. The fuel is supplied using an annular fuel injecting ring, as shown in Figure 3.2. The annular injecting ring comprises multiple injectors located on the circumference of the ring. There can be double annular combustion chamber, as per the requirement. It is thermally efficient, light-weight and compact in size or small in length.

Annular type combustion chamber is mostly used in fighter aircraft, where rapid combustion is required. This type of combustion chamber has very high volume to area ratio compared to other, which helps in having rapid combustion with fewer flow losses. In this type of combustion chamber, maintaining the uniform heat distribution is very difficult.

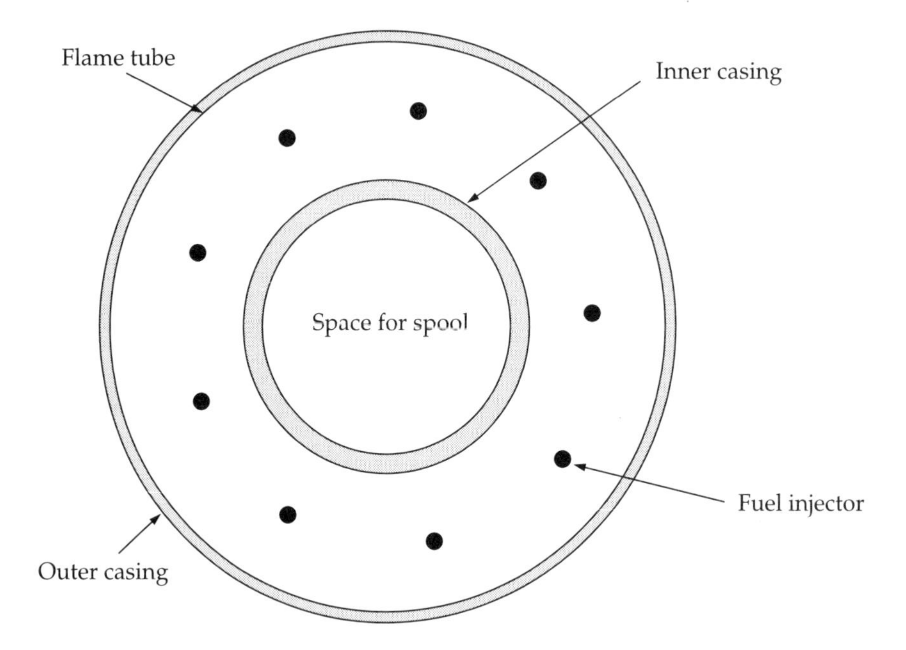

Figure 3.2 Annular type combustion chamber.

3.6.3 Can-Annular Type Combustion Chamber

The can-annular type combustion chamber is a combination of can type and annular type of combustion chamber. The can type and cannular type combustion chambers look alike, but not same, as there is an interconnectivity between the individual can in cannular type, which is not there in the can type, as shown in Figure 3.3. Cannular type of combustion chamber has separate cans to carry and hold the combustion flame with the common annular outer ring. There is an individual fuel injector for each combustion chamber connected to the common annular ring. Primary air is supplied similar to can type combustion chamber, but the individual secondary air is not fed to the individual combustion chambers like can type.

Each combustor is connected internally to one another in series which helps in conveying the heat from one combustor to another. In this type of combustion chamber, all the combustors have an ignition source to start the combustion process independently. The ignition system is in selected combustion chambers, not in every combustion chamber, hence, the interconnection helps in carrying the combustion flame in all the combustion chambers. Can-annular type combustor is mostly used in transport aircraft.

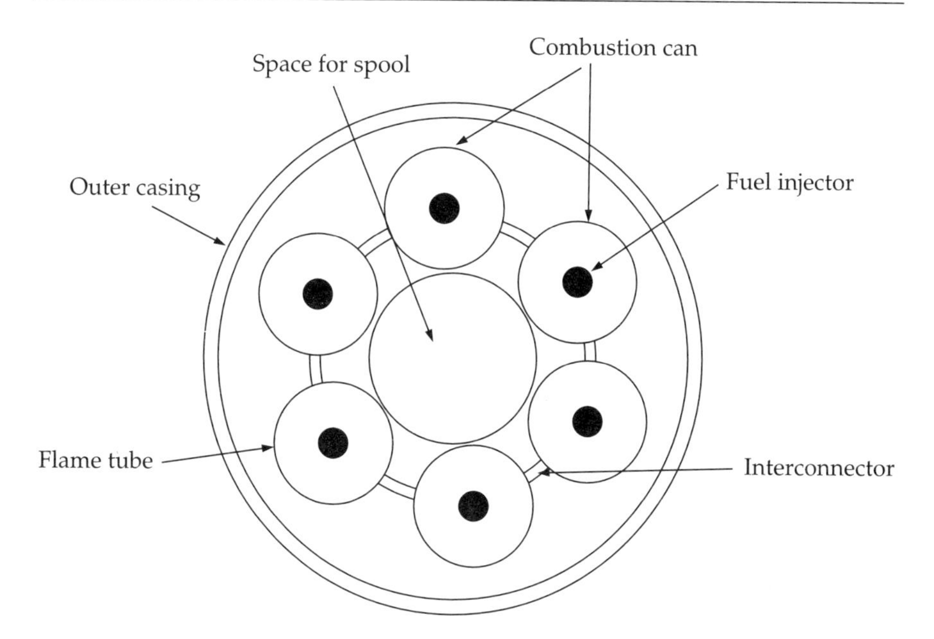

Figure 3.3 Can-annular type combustion chamber.

3.7 COMBUSTION CHAMBER GEOMETRY

With reference to the structure, the leading part of the combustion chamber is snout, which helps the flow to bifurcate or split into primary and secondary supply in a designed ratio. The secondary air flows around the combustor, whereas the primary air flows through the combustor. After the snout, the swirl vanes are located, which gives turbulent effect to the primary air. Swirl vanes are located around the fuel injector so that the turbulent primary air coming out from the swirl vanes or swirlers mixes with the fuel and atomises it, as shown in Figure 3.4.

The stoichiometric air-fuel ratio for most of the hydrocarbons is 15:1. With an overall air-fuel ratio of 60:1 in a gas turbine, only about a quarter of the total air must be admitted to the reaction process, whereas the rest of the air is admitted in reducing the gas temperature to meet the turbine inlet temperature requirement. This reaction zone is called *primary zone*.

The primary zone should have the flame stabiliser such as baffles or swirlers vane to establish a recirculation zone. The stability parameters indicate that it is better to provide a small number of large baffles than a large number of small baffles. Because of the problem of mixing air and fuel followed by combustion initiation, some amount of secondary air is introduced which ensures adequate oxygen with the fuel for complete combustion downstream of the nominal reaction zone. This zone is called *secondary zone*. Basically, nickel-based alloy and ceramic composites are used for fabricating the combustor.

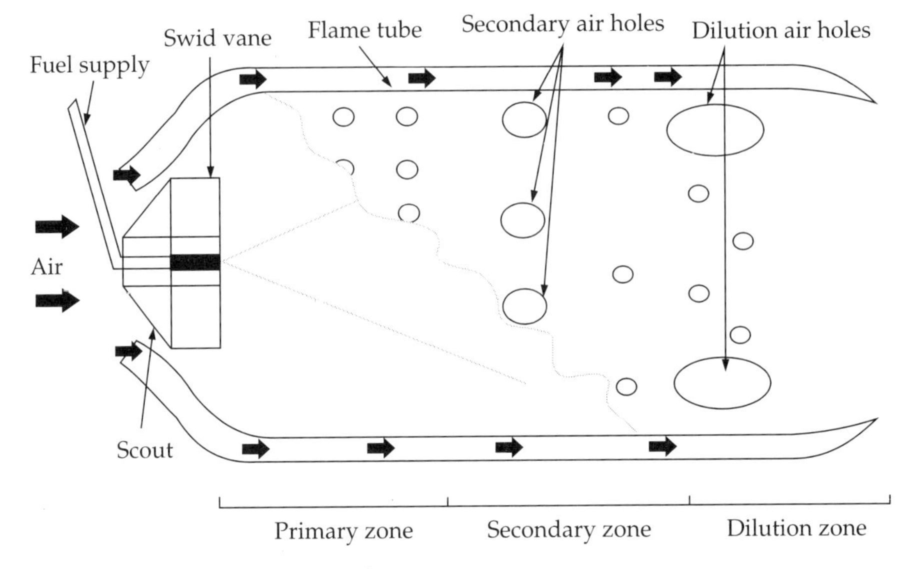

Figure 3.4 Combustion chamber geometry.

3.8 COMBUSTION STABILITY AND INSTABILITY

Combustion stability means smooth burning. It can also be defined as the ability of a combustor to remain alight over a wide range of operation. But, there are some factors which make the combustion unstable, which are as follows:

1. Local and instantaneous air-fuel ratio: For varying fuel supply as per the demand, the fuel-air ratios vary, thus affecting the stability part of the combustion chamber.

2. Instantaneous pressure or pressure fluctuations: For a given combustor cross section, the injected atomised fuel may collide and combine with each other and gets decelerated near wall of combustion chamber. This leads to increase in pressure, which results in instantaneous pressure change in combustion chamber.

3. Instantaneous temperature-based or heat release rate: It purely depends on instantaneous pressure.

Main causes of combustion instabilities

Combustion instabilities may arise due to the following causes:

1. Highly turbulent mixing
2. Variation in combustion flame area, i.e., non-uniform combustion area
3. Variation in the compressor exit flow
4. Vortex formation around the flame holder

Flame holder

The *flame holder* is a component in a combustion chamber located around the fuel injector which helps in deceleration of the high-velocity air entering the combustion chamber so that the combustion flame is stabilised. Flame holder is a dead body in the flow field, which matches the high-velocity air and fuel mixture with the flame velocity. Mesh screens or perforated plates and solid or hollow bodies are used as flame holders.

Flame tube cooling

The flame tube receives energy by convective heat transfer from the hot gases and flame. A common practice is to have small annular gaps between overlapping sections of the flame tube. A film of cool air insulates the surface from hot gases.

3.9 FUEL INJECTION SYSTEM

Fuel injection system is a very important component of a combustor, which plays a major role in the combustion process. The fuel injector injects the fuel at proper velocity, angle and as a fine spray.

The fuel injector exit area decides the velocity and cone angle of injection. Surrounding the injector there is an air supplying injector, which mixes with the fuel and also atomises. The fuel atomisation or presence of two-phase mixture is decided by the injector design and the downstream pressure.

The injectors are classified based on the type of mixing, i.e., premixed or diffusion type injectors, and also on the number of injection cone, i.e., simplex or duplex injector.

Simplex injector

A simplex fuel injector injects the fuel from the swirl chamber. The fuel inside the injector gets swirl due to the tangential entry of air into it. The flow is decided by single nozzle flow control mechanism. The swirling effect makes the fuel mix properly with the air and forms fine droplets after the exit of the fuel injector. These fine droplets with the air are injected into the combustion chamber at right cone angle and at a calculated distance. In this type of injector, we can see only one cone which can be varied in a way such that, its cone angle, diffusion rate and cone length can be varied by changing the exit place of the individual injector.

Duplex injector

In the type of fuel injector, there are three injectors or nozzles one inside the other. The fuel is injected in two different fashions—the primary part of the fuel is supplied from the centre or core of injector and the main part of the fuel is supplied from the periphery or from the intermediate injector. Air is supplied at the outermost injector casing, as shown in Figure 3.5.

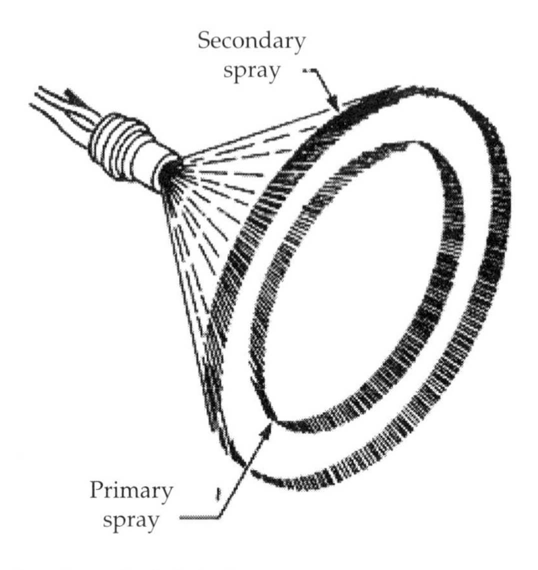

Figure 3.5 Duplex type fuel injector.

While takeoff time, both the fuel injecting elements (primary and secondary) are used, whereas in cruise condition, the main part of the fuel supply is switched off. The fuel injecting nozzle creates concentric jets of conical shapes. High pressure turbulent and vertical flow zones create droplets. The conical injection shape increases the surface area which aids in evaporation and air mixing. The air jet has one more job, i.e., avoiding the deposition of combustion products on the fuel injector, which leads to fuel choking.

3.10 NOZZLE

The nozzle is a thrust producing duct located after turbine section in a gas turbine engine or it is the end component in any propulsion system. It is also called *exhaust nozzle*. The primary purpose of the nozzle is to transform the pressure energy into kinetic energy of combusted gas to produce thrust. It also collects the flows coming from different paths and straightens the flow of exhaust. For a large value of a specific thrust, the kinetic energy of exhaust must be very high, nothing but high-velocity exhaust jet. The expansion process in the nozzle is controlled by the upstream pressure ratio. The maximum thrust for a given mass flow rate is obtained mainly when the exit pressure P_e from the nozzle is equal to the ambient or atmospheric pressure P_0, which is nothing but the complete expansion of gas.

Functions of nozzle

Nozzle performs the following functions:

1. It accelerates the hot flow.
2. It matches the flow pressure to the atmosphere pressure as close as possible, which increases the specific thrust.
3. It permits afterburner operation located in between turbine and nozzle without affecting the main engine process.
4. It allows the inner wall cooling by using bypass air.
5. It mixes the bypass stream with the core stream of engine to decrease the noise and flow eddies.
6. It allows thrust reversing by adopting the clamshell type thrust reversal without affecting the internal flow.
7. It allows the noise and infrared radiation suppressing techniques.
8. Thrust vectoring system can also be integrated with the nozzle.

All the above functions should be attained at a minimum cost, weight factor and bow tail drag. The exhaust nozzle has mainly three parts, namely, exhaust cone, tail pipe and exhaust nozzle. The exhaust cone has an outer duct, inner cone and struts. Nickel alloy, titanium alloy and ceramics are used as material for making nozzles.

The nozzle is used to direct the turbine discharge aft gases to the atmosphere at maximum velocity to produce the required thrust by expanding the gases. The gases must expand completely and discharge vortex-free and axial flow. Heat and pressure energy are converted into velocity, and hence, into thrust. If the gas reaches sonic speed, the flow will be choked and there will be no further expansion unless the gas temperature is increased or throat area is increased or upstream gas pressure is increased. The flow inside the nozzle can also be called *Fanno flow*. It can be considered that the flow inside the constant cross section nozzle with friction and without heat exchange is a Fanno flow.

In nozzle, it is isentropic expansion process (no heat interaction). As Fanno flow equation is applicable to the nozzle, the maximum entropy is found at Mach 1 and there is no change in heat and work. Hence, the stagnation temperature also remains constant throughout the nozzle.

Inside nozzle, there is a thin insulating wall which protects the inner surface from excessive heating and also avoids maximum amount of heat exchange; meanwhile, the flow tries to have choked condition. At high speeds, it is necessary to use both continuously variable intake and nozzle capable of continuous variation of nozzle exit area. The nozzle is designed to match high-pressure high-temperature flow with the atmospheric pressure. The ratio of nozzle entry pressure to the nozzle exit pressure is called *nozzle pressure ratio*.

3.11 TYPES OF NOZZLE

There are mainly two types of nozzle, namely, convergent and convergent divergent, which are used in almost all the types of air-breathing propulsion systems.

3.11.1 Convergent Nozzle

The convergent nozzle is a simple convergent duct or gradually decreasing area duct, which accelerates the flow by expanding the high-pressure gas, as shown in Figure 3.6. It is also called *subsonic nozzle*. When nozzle exit pressure ratio is low (less than 1.86), the convergent type of nozzle is used. *Exit pressure ratio* is the ratio of flow pressure at the exit of the nozzle to the atmosphere pressure. The convergent shape of the nozzle increases the velocity of the gas as the cross section of the duct or nozzle reduces gradually. It is generally used for subsonic aircraft engines where the momentum thrust is more with less or almost no pressure thrust.

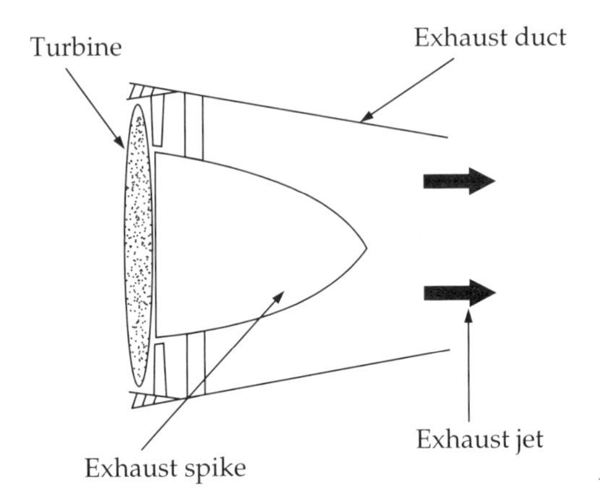

Figure 3.6 Convergent type nozzle.

3.11.2 Convergent-divergent Nozzle

The convergent-divergent nozzle is used exclusive for higher supersonic or hypersonic jets or to expand the gas to supersonic or hypersonic speed. It is also called *supersonic nozzle*. It has a convergent section, a throat and a divergent section, as shown in Figure 3.7. The section of the nozzle where the cross-sectional area is very less is called *throat*, which is located in between convergent and divergent section. The convergent-divergent nozzle is used when the nozzle exit pressure ratio is high (more than 1.86). If the nozzle pressure ratio is less than that, still this nozzle can be used, but the weight and length of the engine increases which is like dead weight of no use.

Convergent-divergent nozzle and convergent nozzle incorporate variable geometry system, thrust vectoring, as shown in Figure 3.8, and other aerodynamics features.

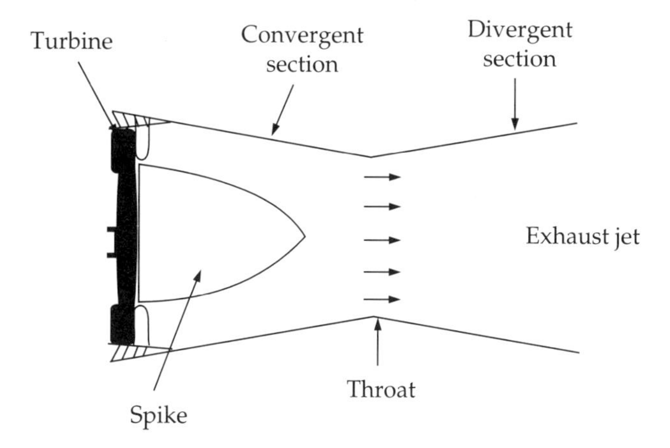

Figure 3.7 Convergent-divergent nozzle.

Figure 3.8 Thrust vectoring.

The convergent-divergent nozzle shows less range and less efficiency if operated outside its design point or envelope. Hence, to compensate this, these nozzles are designed as variable area nozzle also. The convergent-divergent nozzle works in three major conditions, namely, critical, subcritical and supercritical conditions. These conditions are based on shock wave location, which is nothing but choked condition. If the shock is created inside the convergent section near the throat, then it is called *subcritical condition.* If the shock is exactly in the throat section, then it is called *critical condition,* and if the shock is at divergent section, then it is called *supercritical condition.* The main drive to adopt the convergent-divergent nozzle is to match the exit pressure with the ambient so that the pressure thrust term can be eliminated and thrust value with its efficiency can be increased.

3.12 THROAT CONDITIONS FOR SUPERSONIC NOZZLE

1. The nozzle should choke at the throat (the mass flow at a given temperature through the nozzle is sufficient to reach sonic speeds at the smaller cross-sectional area).
2. By the time the flow come to the exit area of the convergent section, the flow should be transformed from subsonic speed to sonic speed.
3. The entry pressure to the nozzle has to be significantly much above the ambient condition at all times.
4. The pressure at the exit of the nozzle should be equal to or more than an ambient condition to avoid reverse flow.

As the gas enters the nozzle from the turbine at high pressure and temperature, which is at subsonic velocity it is forced to accelerate in the convergent section. The flow reaches a sonic velocity at the throat. This leads to shock at the throat which separates the flow into the subsonic region and supersonic region. From the throat section, the nozzle diverges (divergent section), which leads to more expansion of gas and the velocity progresses to supersonic velocity.

3.13 NOZZLE EFFICIENCY

Nozzle efficiency can be defined as the efficiency with which a nozzle converts the pressure energy into kinetic energy or it is the ratio of actual change in kinetic energy to the ideal change in kinetic energy at a given pressure ratio. It can also be defined as the ratio of energy converted into kinetic energy to the total potential energy which could be converted into kinetic energy [Eq. (3.8)].

$$\text{Total energy, } E = h_0 - h_{\text{exit 2}} \tag{3.6}$$

$$\text{Actual energy, } E = h_0 - h_{\text{exit}} \tag{3.7}$$

$$\text{Nozzle efficiency, } \eta = \frac{(h_0 - h_{\text{exit}})}{(h_0 - h_{\text{exit 2}})} = \frac{(U_{\text{actual}})^2}{(U_{\text{ideal}})^2} = v_e^2 \tag{3.8}$$

where v_e is velocity coefficient, h is enthalpy, U is peripheral velocity, h_0 is the stagnation enthalpy, h_{exit} is the actual enthalpy at the exit of nozzle, h_{exit2} is the ideal enthalpy at the exit of nozzle.

To calculate the exit jet velocity using the pressure ratio and temperature, Eq. (3.9) can be used.

$$\text{Jet velocity } v_j = \sqrt{\frac{\gamma}{\gamma-1} RT_e \left(1 - \left(\frac{P_a}{P_e}\right)^{\frac{\gamma-1}{\gamma}}\right.} \tag{3.9}$$

where, γ is specific heat ratio, R is gas constant, T_e is nozzle exit temperature, P_e is nozzle exit pressure, and P_a is atmosphere pressure.

3.14 UNDEREXPANDING AND OVEREXPANDING NOZZLES

These are the nozzles which are classified based on the flow expansion process or based on the exit pressure of the nozzle. The underexpansion type of nozzle has high pressure at the exit section of the nozzle as compared to the ambient pressure, as shown in Figure 3.9. The underexpansion nozzle means the flow inside the nozzle does not expand completely to match with the ambient condition. This type of nozzle leads to expansion fan at the nozzle exit surface with high-temperature shock at the boundary layer.

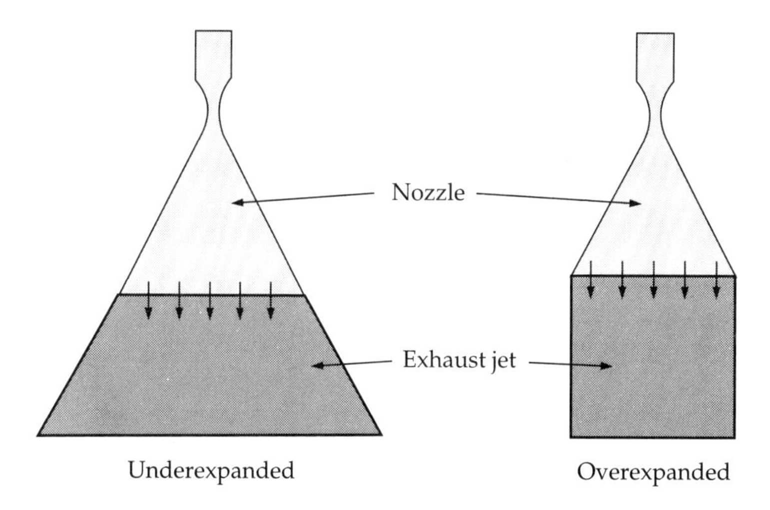

Figure 3.9 Underexpanding and over expanding types of nozzle.

The overexpanding nozzle has low pressure at the exit of the nozzle as compared to the ambient condition. In this type of nozzle, the flow expands more because of some factors like less reduction in mass flow as compared to design point, variation in altitude, etc. This leads to flow separation before the flow exits from the nozzle, which in turn leads to side force on inside wall of the nozzle. This produces oblique shock wave reflections at the exit, which reduces the efficiency with its high-pressure region.

Interaction of nozzle flow with adjacent surface

1. If the exit pressure is same as the ambient pressure, then the flow coming out will be laminar with less load on the nozzle wall.
2. If the exit pressure is greater than the ambient pressure, then it will produce expansion fan and affect the nozzle exit tip.
3. If the exit pressure is smaller than the ambient pressure, then it will lead to oblique shock wave and flow separation near the tip of the nozzle.
4. The flow separation and shock wave lead to high pressure at upstream, and hence, side force on the nozzle.

3.15 VARIABLE AREA NOZZLE

A complete expansion of gas or almost no pressure thrust and improved fuel consumption can be achieved by varying the exit area of the nozzle. A change in nozzle exit area at takeoff and approach can increase or reduce the jet velocity followed by some amount of noise variation.

In variable area nozzle, the nozzle petals have got some degree of freedom, whereas in normal nozzle, it looks like a duct with no petal-like structure and movement. These nozzle petals are brought into motion using hydraulic actuators which deploy at particular flight conditions on pilot inputs. In a convergent nozzle, as the requirement of thrust varies, the nozzle petals are deployed accordingly. If high amount thrust is required then the nozzle exit area is decreased and if thrust required is low amount of thrust is required then exit area is increased. This condition is true for engine without afterburner case and it is opposite when engine is with afterburner. From the ideal condition of the nozzle exit area, the area can be increased or decreased also. In other condition, throat area can also be varied by moving the translating cone inside or outside the nozzle aperture.

In thrust augmentation, i.e., afterburner, the exhaust gas temperature crosses the allowable limit and by increasing the exit area of the nozzle, the temperature of the gas can be brought within the limit. If thrust augmentation is used, it means the requirement of thrust is more and the temperature and pressure of the flow are increasing. Hence, to meet the main function of the nozzle, i.e., flow expansion and acceleration and to make the exit pressure equal to the ambient temperature, the exit area of variable area nozzle is increased.

In variable area nozzle, there must be precise synchronisation between the reheat actuation switch and variable area nozzle deploying system. If there is a significant delay in the process, then it leads to surging in any set of the compressor, normally in high-pressure compressor set.

3.16 THRUST REVERSAL

Thrust reversal is a mechanism to reverse the direction of thrust force generated by the engine. By changing the thrust force direction, the momentum of an aircraft can be demolished or drag can be increased or in simple words, we can apply brakes. This mechanism is actuated mainly after aircraft touchdown.

The major classifications of thrust reversal mechanism are cascade type and clamshell type. The *cascade type thrust reversal* is one where the compressed air from the compressor is blocked by getting inside the combustion chamber using internal block door and pushed out from the passage provided in between the compressor and the combustion chamber stage externally. While doing this process the cowling near that section opens by sliding pneumatically. This is also called *cold flow thrust reversal.*

The *clamshell* is another type of thrust reversal, where the two bucket-like structures are provided in the aft portion of the nozzle. These structures are attached to the nozzle externally in the initial condition and deployed on the requirement. Clamshell thrust reversal is pneumatically operated using the engine bleed air after the compressor, which is discussed in the previous chapters. It is also called *bucket type thrust reversal* or *hot stream thrust reversal*. The major disadvantage of this type is that the structure is in direct contact with the hot stream, which leads to structural damage.

IMPORTANT QUESTIONS

1. Define combustion process. Name the three major components required for combustion process.
2. Define Rayleigh flow. Is there any temperature change in Rayleigh flow?
3. In gas turbine combustor, which are the components that make the flow into two parts and give swirling effect?
4. Explain the classification of air and its function in combustion process.
5. What do you mean by stoichiometric mixture and equivalence ratio?

6. List the factors which affect the following: (a) Combustion chamber design, (b) Combustion chamber performance.
7. Explain the combustion process in detail.
8. Write the differences between can type, annular type and can-annular type combustion chambers.
9. Explain the two different types of fuel injectors.
10. Define nozzle. Write functions of nozzle.
11. Define convergent nozzle. Explain its working. What will be the nozzle exit condition when it is operating at design condition? Discuss.
12. Can convergent nozzle be used in supersonic aircraft? Justify your answer.
13. Write the difference between convergent nozzle and convergent-divergent nozzle.
14. Explain the three different working conditions of convergent-divergent nozzle.
15. What do you understand by underexpanding nozzle and over-expanding nozzle?
16. Explain variable area nozzle and thrust reversal. Which type of thrust reversal has better performance? Discuss.

CHAPTER 4

TURBINES

OVERVIEW

In this chapter, we will study about the turbine, its classification, working principle, efficiency, and so on. We will also look into some concepts like velocity triangle, blade shape and some performance parameters. By the end of this chapter, we will find answers to some questions related to the working of turbines, selection of a turbine for a particular engine and the working of multi-stage turbines.

Turbine is the main rotating part of a gas turbine engine, which converts the pressure and heat energy into kinetic and mechanical energy. Turbines transform heat energy into mechanical energy, i.e., high-temperature high-pressure flow into high-velocity turbine rotation by making the fluid to flow through the turbine, and during this transformation, the pressurised fluid expands in the turbine, leading to increase in the flow velocity at the downstream of the turbine.

Turbine works on momentum principle, i.e., the momentum of the gas creates change in the angular momentum of the turbine. During the flow expansion, the heat and high-pressure energy of the gas are converted into kinetic energy of turbine and kinetic energy of the flow across the turbine. The momentum imparted by the gas on turbine blades makes the turbine rotate with its spool or shaft.

The turbine blades are designed based on the requirements of individual turbofan, turboshaft or turbojet engines. In turboshaft or turboprop engine, the power extraction is high and power is extracted using free or power turbine located at the end of the turbine stage. 85% of the total engine power is used to drive the propeller or power shaft in a turboprop or turboshaft engine. Turbines extract the power required to drive the compressor and necessary accessories for continuous operation. The requirement of turbine stages is determined by the amount of energy to be extracted from gas flow to produce necessary shaft horsepower.

Counter-rotating turbine is one where the turbulence created by the upstream turbine is eliminated by a downstream turbine which rotates in opposite direction to the upstream turbine and also leads to maximum utilisation of energy.

The two primary parts of turbine are:

1. Stator nozzle
2. Rotor blade

As discussed in the compressor component section 2.5, the turbine has two major components, as mentioned above. One is stationary component which guides and accelerates the flow whereas another is rotary component which extracts the energy from the gas flow. The rotating disk with blades extracts the energy when the flow impinges on the blades and gives a momentum effect. The stator vanes on the casing disk are in a convergent shape which increases the flow velocity and directs the flow so that it impinges the rotor blades at right angle and velocity.

The turbines are classified as:

1. Impulse type turbine
2. Reaction type turbine

An impulse type turbine stage is characterised by the expansion of gas, which takes place only in the stator nozzles. It means the energy transformation is in the stator, which is usually convergent in shape. The decrease in pressure energy can be seen only in stator part. The rotor blades are just energy absorber where no energy transformation takes place, as shown in Figure 4.1. Further, the rotor blades convert the kinetic energy of gas into mechanical work by changing the momentum of the gas more or less at constant pressure. The relative eddy causes the flow in the impeller passage to deviate from the blade angle, resulting in the shifting of the apex of actual velocity triangles at the exit called *slip* and *slip velocity*. In simple words, if the magnitude of the relative velocity at inlet and exit of turbine does not change, then it is called *impulse type turbine*, and if it changes, then it is called *reaction turbine*.

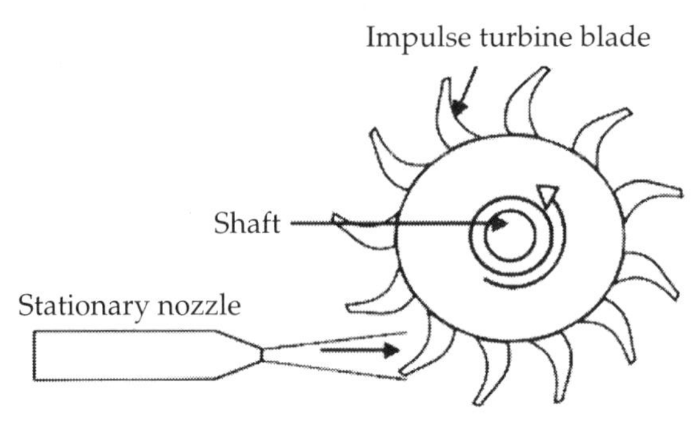

Figure 4.1 Impulse turbine.

A *reaction type turbine* is one in which expansion of the high-pressure gas into high velocity takes place both in the stator and the rotor stages, which is nothing but the energy transformation in both rotor and stator stages. The main functions of the rotor are same as that in the impulse type turbine, which are described as follows:

1. The rotor converts the kinetic energy of the gas into mechanical work.
2. Energy transforms from pressure energy to kinetic energy of the gas.
3. Contributes a reaction force on the rotor blade.

The reaction force generated is due to the increase in the velocity of the gas relatively to the blade velocity. This results from the expansion of the gas during its flow through the rotor, as shown in Figure 4.2.

Figure 4.2 Reaction turbine.

4.1 OPERATING PRINCIPLE OF IMPULSE TURBINE

Impulse turbines are the turbomachines in which there is no change in static pressure of the fluid inside the rotor. The rotor blades cause only energy transfer and no energy transformation. The energy transformation with energy transfer takes place in fixed (stator) blades or nozzles only, as shown in Figure 4.3(a).

The movement of turbine rotor is only because of impulsive action of the fluid on the rotor. No acceleration is added to the fluid in the rotor. Hence, boundary layer separation is more on the blade surface, which leads to loss and lower stage efficiency. Pelton wheel and paddle wheel are the examples of this type of turbine. The impulsive action on the rotating blades is due to the absolute velocity component.

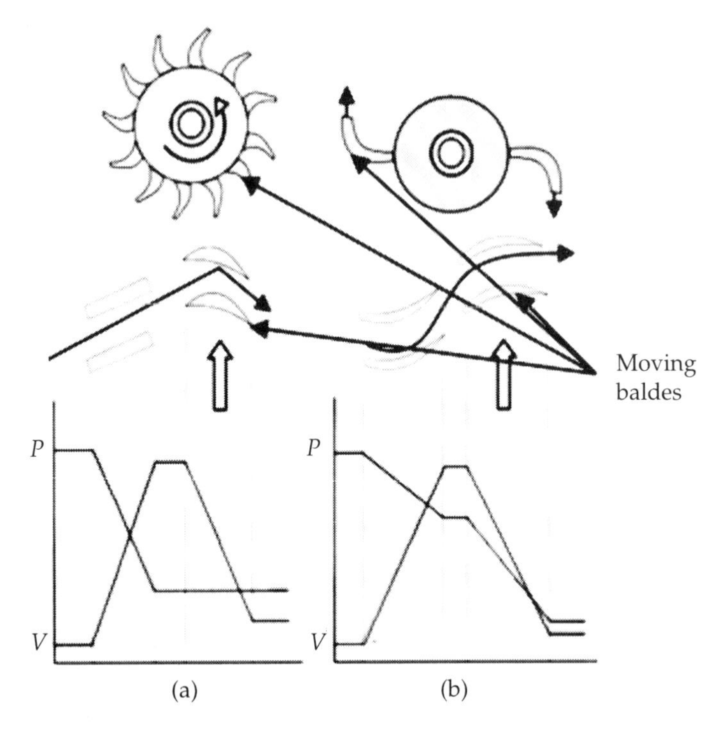

Figure 4.3 Pressure and velocity distribution for (a) impulse and (b) Reaction type turbine.

4.2 OPERATING PRINCIPLE OF REACTION TURBINE

In this type of turbine, the change in static pressure occurs in both parts of the turbine (rotor blades and stator blades), as shown in Figure 4.3(b). The *degree of reaction* of turbomachine stage is defined as the ratio of static pressure head change in the rotor to the static pressure head change across the stage. The rotor experiences both energy transfer and energy transformation. As the flow expands in the rotor stage with the acceleration in the flow, the boundary layer separation becomes less, hence, the loss is less. Therefore, the reaction type of turbines are considered to be more efficient. This is mainly due to the continuous acceleration of flow with less loss.

4.3 VELOCITY TRIANGLE

The flow magnitude at entry and exit of turbomachine stage in all the possible directions can be described using vectors in a triangular shape. As per the angle created by the flow vectors, such triangles are called *velocity triangle*. The velocity triangle gives the magnitude and direction of each flow component and the angle between them. This helps in studying and improving the performance parameters like work done, efficiency and many other parameters of the flow. The vector describes the velocity of the flow which is used to find the remaining parameters of each flow component.

The components of velocity triangle are:

1. Peripheral velocity denoted by u or U
2. Relative velocity denoted by w or V_r
3. Absolute velocity denoted by c or v

Peripheral velocity is the velocity of the turbomachine measured at the tangential section or peripheral section. The reference point for the velocity component is the initial velocity of the turbomachine or its rest position. Mathematically, the peripheral velocity is the product of the angular momentum of the turbomachine and the radius of the turbomachine. *Relative velocity* is the velocity that relates the fluid velocity with the turbomachine rotor velocity. The relative velocity is the velocity where within the frame, the blade appears to be stationary. The absolute velocity is the velocity of the fluid which is flowing across the turbomachine as shown in Figure 4.4. The reference point for this velocity component is the stator part of the turbomachine or the velocity of the fluid when it is at rest, which is nothing but zero. The peripheral velocity and relative velocity components are derived from the absolute velocity component. In designing the turbomachines, these velocity components and triangle are used, which explain the torque produced and the power of turbomachine.

The performance of a turbomachine is characterised by various dimensionless parameters like loading coefficient and flow coefficient. The loading coefficients explain how much a turbomachine is being loaded in getting the required outputs and flow coefficients explains how much the flow is efficient in getting the required output from the turbomachine.

Blade efficiency is also known as *utilisation factor*. The blades which are designed to produce the expected output also have some efficiency, which is not 100% and that decides how much energy is being utilised from the fluid flowing through the turbomachine. Even though in reaction turbine, the blade work and stage work are very similar in terms of energy transformation, the blade and stage efficiency need not to be same. Blade work is the sum of various kinetic energies.

As shown in Figure 4.4, the individual components are again classified based on the flow direction with respect to blade position, i.e., each velocity component is having two more subcomponents, namely, an axial component and a tangential component and the resultant of these components is the actual velocity component.

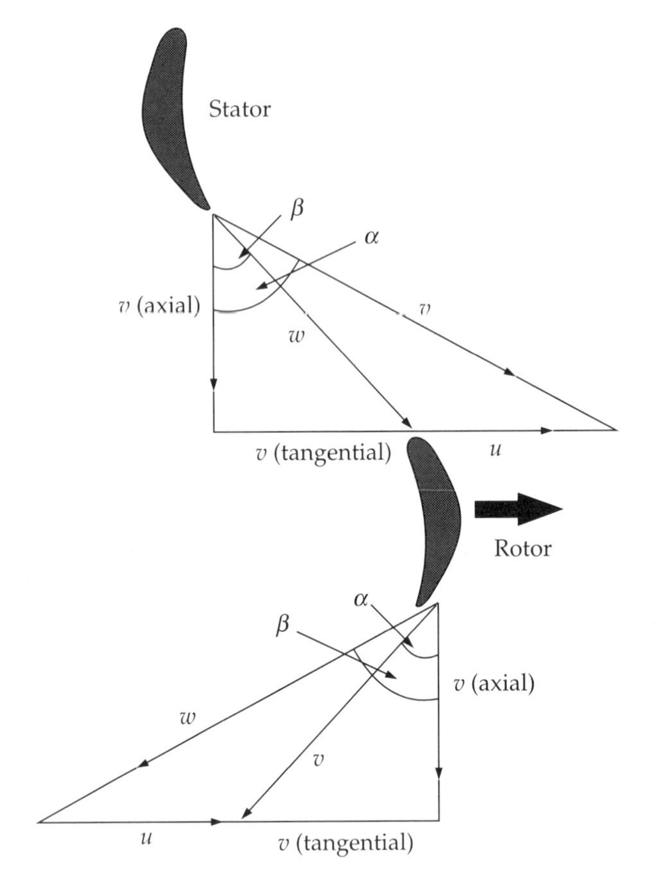

Figure 4.4 Velocity triangle of axial type turbine.

The angle between the axial component of absolute velocity and the absolute velocity is known as *absolute angle,* which is, denoted by α and the angle between the axial component of absolute velocity and the relative velocity is known as *relative angle* or *blade angle,* which is denoted by β. These angles are used to find the unknown velocity components and all the components together are used to find the required parameters of the turbomachine.

The performance of a turbomachine is characterised by various dimensionless parameters like loading coefficient and flow coefficients as discussed before, which are given by

Loading coefficient, $\varphi = \dfrac{W}{u^2}$, where W is the work done and u is the peripheral velocity.

Flow coefficient, $\phi = \dfrac{C_a}{u}$, where C_a is the axial component of absolute velocity.

The relation between loading coefficient and flow coefficient is given by Eq. (4.1).

$$\varphi = \phi(\tan \alpha_2 + \tan \alpha_3) = \phi(\tan \beta_2 + \tan \beta_3) \tag{4.1}$$

Blade efficiency or utilisation factor is given by Eq (4.2).

$$\eta_b = \frac{\text{Rotor blade work}}{\text{Energy supplied}} = \frac{W}{E_{rb}} \tag{4.2}$$

where the work done [Eq. (4.3)] and the energy [Eq. (4.4)] supplied to the turbine are written in terms of flow velocity components

$$W = \dot{m}\left[\frac{(v_1^2 - v_2^2)}{2} + \frac{(u_1^2 - u_2^2)}{2} + \frac{(w_2^2 - w_1^2)}{2}\right] \tag{4.3}$$

$$E_{rb} = \dot{m}\left[\frac{(v_1^2)}{2} + \frac{(u_1^2 - u_2^2)}{2} + \frac{(w_2^2 - w_1^2)}{2}\right] \tag{4.4}$$

where $\dot{m}$ is the mass flow rate of fluid.

The isentropic efficiency of the radial turbine is less than the isentropic efficiency of the axial turbine. In the calculation of rotor specification, the relative velocity is considered and in the calculation of stator specifications, absolute velocity is considered. The polytrophic efficiency of each stage is less than the isentropic efficiency or complete turbine stage.

The work done and efficiency can also be expressed in Eqs. (4.5) and (4.6).

$$W = \dot{m}_g C_P (T'_{03} - T_{04}) \tag{4.5}$$

$$\eta_T = \frac{T_{03} - T_{04}}{T_{03} - T'_{04}} \tag{4.6}$$

where $\dot{m}_g$ is the mass flow rate of the gas, C_P is the specific heat at constant pressure of the gas, T_{03} is the turbine inlet temperature of the gas, T_{04} is the turbine exit temperature of the gas and T'_{04} is the ideal exit temperature of the turbine.

The work done by the turbine [Eq. (4.7)] can also be defined as the amount of work done to drive the compressor efficiently.

$$W_T = \frac{W_C}{\eta_m} \tag{4.7}$$

where W_T is the work done on the turbine, W_C is the work done by the compressor and η_m is the mechanical efficiency.

4.4 BLADES

In order to understand the actual energy transfer process and flow through the blades, we use velocity triangle. Based on the relative angle, or in general, blade angle β, the blade shapes are classified as forward curved, backward curved and radial curved in radial or centrifugal type turbine. If this relative angle or blade angle is acute ($\beta < 90°$), then such blade is called *backward curved blade*. If the relative velocity component is perpendicular to the axial component of absolute velocity ($\beta = 90°$), then it is called *radial curved blade,* and if the relative angle is obtuse ($\beta > 90°$), then it is called *forward curved blade,* as shown in Figure 4.5. The radial curved blade gives a good performance, backward curved blade gives maximum efficiency and stability and the forward curved blade gives maximum pressure ratio for a given tip speed.

Turbine blades are exposed to very high temperature and pressure, which leads to thermal expansion and thermal wear. To overcome this problem, the turbine blades are cooled continuously using relative cooler air. The cooling air is taken from the last stage of compressor which is at comparatively lower temperature. If the air is taken from the first few stages to cool the turbine blades, then the reduction in mass flow will affect the compressor stage.

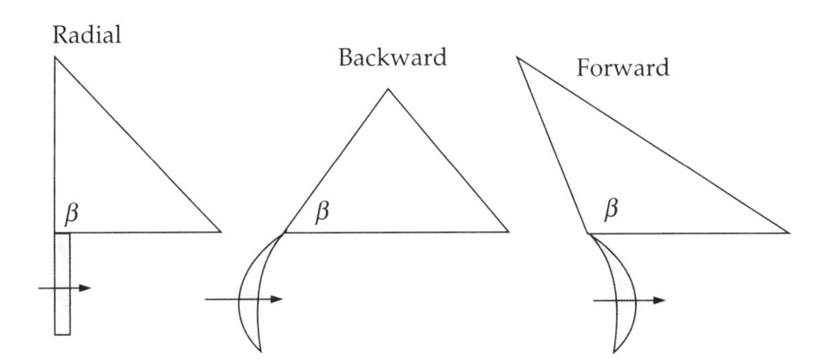

Figure 4.5 Blade shapes.

Thermal expansion of inner and outer layer of turbine blade may change the shape of the turbine blade also. Thermal barrier coating is also done on the turbine blade to avoid the thermal expansion. Usually, nickel-based alloy with chromium and cobalt is used as blade material.

4.5 VELOCITY COMPOUNDING OF MULTISTAGE IMPULSE TURBINE

When a flow expansion to be achieved in an impulse turbine is large, it is not possible to achieve it in just one stage. If this has to be achieved in a single stage only, then this will lead to either a large diameter turbine or a very high rotating speed turbine.

A single stage turbine will transform a large amount of pressure energy into kinetic energy, which leads to high (supersonic) peripheral velocity of rotor. This is not acceptable as the efficiency will be very less. In a multi-stage expansion of impulse, turbine has to generate high velocity of the fluid or utilise the pressure energy to the maximum. This can be accomplished by expanding the large-pressure gas in a nozzle row. The expansion leads to rise in flow velocity after stator nozzle. The high-velocity flow is made to flow across the stator and impinge on the rotor blades. This decreases the absolute velocity due to energy transfer to the moving blades. Since the turbines are of impulse type, the pressure of fluid remains almost constant after expansion in rotor blade row. This cycle is repeated to utilise the high velocity for driving the rotor. Such stage is called *velocity stage*.

4.6 PRESSURE COMPOUNDING IN MULTISTAGE IMPULSE TURBINE

Consider the upstream pressure to the turbine is large and to use the pressure energy, the turbine nozzle should be of convergent-divergent type for high velocity, assuming the maximum expansion of gas. This leads to more expensive and complicated design. At the stator nozzle exit, there can be losses and shocks also. For avoiding all these problems, the total pressure at upstream of the turbine is expanded in multiple stages. This is called *pressure compounding* of impulse turbine. This method is generally used where the energy transformation takes place only in the stator and energy transfer in the rotor.

4.7 REACTION TURBINE

If the degree of reaction is unity, then the turbine is called *pure reaction turbine*. Leakage is a serious problem in the reaction turbine blade in the high-pressure section where the blades are short and percentage of tip clearance area is relatively large compared to the total area. Leakage depends on the clearance, velocity ratio and stage pressure ratio. Hence, there is a sealing problem at blade tip due to the existence of pressure gradient.

Turbine tip clearance of 1.52 mm to 1.02 mm is maintained for non-shrouded and for shrouded type of turbines. Multi-stage reaction turbines have a large amount of leakage, which is divided into smaller values in individual stages. Thus, the reaction stages are like the pressure compounding stage, with a new element of reaction introduced in them, i.e., accelerating the flow through the blade rows also.

4.8 DEGREE OF REACTION

The *degree of reaction* can be defined as the ratio of isentropic change of enthalpy in the rotor to the isentropic change of enthalpy in the stage. It can also be expressed in terms of pressure or velocity or enthalpy or flow geometry of a stage.

Zero degree of reaction

In a single stage impulse turbine, expansion of the gas occurs in the nozzle blade rows and its thermodynamic state remains unchanged between the stations. Therefore, the flow angles, static pressure and enthalpy at the entry and exit of the rotor in the stage are the same. Performance is represented in terms of loading coefficients and flow coefficients, i.e., $\varphi = 2$.

50% reaction turbine

The flow and cascade geometries in the fixed and moving blade rows are such that they provide equal enthalpy drops. This does not imply that the pressure drops in the two rows are also same. Performance is represented in terms of loading coefficient and flow coefficient, i.e., $\varphi = 2(1 - R)$.

IMPORTANT QUESTIONS

1. How does turbine transform the energy from its one form to another form? Explain the process.
2. With a neat diagram, explain the operating principle of impulse turbine and reaction turbine.
3. What do you mean by energy transfer and energy transformation in turbine?
4. Why do we use velocity triangle? Define the different velocity components, subcomponents and their angles.
5. How does the work done on turbine affect the work done by the compressor? Is isentropic efficiency more than polytrophic efficiency of turbine?
6. How do you classify the blades based on the blade angle?
7. Explain the velocity and pressure compounding in impulse turbine.
8. Write the difference between impulse turbine and reaction turbine.
9. How can a turbine stage or couple of turbine stages drive the complete compressor sets? Justify your answer.

CHAPTER 5

FUNDAMENTALS OF ROCKET PROPULSION

OVERVIEW

This chapter starts with the meaning of rocket followed by rocket engine working principle, thrust and efficiency of a rocket engine, classification of rocket engines with their applications, types of fuel used in the rocket engine, rocket nozzle and types, rocket staging, and so on. By the end of this chapter, we will be able to answer a few questions like how rocket engine is different from other type of engine; when and which type of rocket engine has to be used, etc.

Rocket is a space vehicle, which is used to travel or convey objects beyond the atmosphere region. In simple words, rocket is a vehicle to travel in space. The word 'rocket' sounds like vehicle used to travel in space, but it is not only meant for space. Rockets can be used within the atmosphere and the best example is rocket missiles. Prior to space journey, a rocket must travel through the atmospheric region; hence, the propulsion system of the rocket should be efficient within the atmosphere and beyond the atmosphere.

A *rocket engine* is also known as *rocket motor,* which is a type of non-air breathing engine and propels within the atmosphere and beyond the atmosphere. The rocket engine consumes the oxidiser required for combustion process from the local or onboard oxidiser tank rather than from atmosphere. This phenomenon is not similar to other chemical propulsion systems used within the atmosphere. It means rocket has to carry an oxidiser tank also with the propellant. This adds more weight and volume to the rocket.

5.1 CLASSIFICATION OF ROCKET ENGINE

1. Chemical rocket engine
   (a) Solid fuel motor
   (b) Liquid fuel motor
   (c) Hybrid motor

2. Nuclear rocket engine
   (a) Fusion reaction
   (b) Fission reaction
3. Electrodynamic rocket engine
   (a) Ion and plasma rocket
   (b) Non-ion or Photonic rocket

Chemical rocket engine is one where the fuel (propellant) and oxidiser undergo thermochemical process and produce heat energy, which is used to generate thrust by transforming it into kinetic energy. Based on the physical state of the propellant, it is sub-classified into three types—solid fuel motor, liquid fuel motor and hybrid motor. Fuel cell rocket engine is also another type of chemical rocket engine, which converts chemical energy into electric energy, and hence, produces thrust.

In *nuclear rocket engine*, the fission and fusion process of nuclei generates energy, which is used as thrust. The nuclear rocket engine is a very complex system, which will have many challenges like having a control over the chemical reaction and the heat produced. This type of engine is still conceptual and it is expected in the near future.

In *electrodynamic rocket engine*, electric power is used to excite the ions from the plasma source and those ions are directed using magnetic nozzle to generate thrust. Thrust can also be generated using stimulated laser, which generates photons at very high momentum, which is called *photonic rocket*. The complexity of this system is more and the thrust value is comparatively less. The pollution level can be controlled using electrodynamic rocket engine as compared to chemical rocket engine. In this chapter, we will look into the details of chemical rocket engine only.

5.2 DIFFERENCES BETWEEN AIR-BREATHING ENGINE AND ROCKET ENGINE

Air-breathing engine	*Rocket engine*
1. Altitude limitation	No altitude limitation
2. Thrust decreases with altitude	Thrust increases with altitude
3. Rate of climb decreases with altitude	Rate of climb increases with altitude
4. Flight speed always less than jet speed	Flight speed can be more than jet speed
5. High efficiency and specific impulse	Low efficiency and specific impulse

5.3 PRINCIPLE OF ROCKET PROPULSION

The basic principle of rocket propulsion includes Newton's laws of motion. Newton's third law explains very well about rocket propulsion, which is "every action has equal and opposite reaction". In rocket, high-velocity jet comes out from the engine, which is the action, and forward (opposite to jet) force is produced, which is the reaction. The magnitude of thrust can be determined using the equation of Newton's second law of motion, which is the law of momentum.

Rocket engine in its simplest form consists of a combustion chamber and an expanding nozzle with some accessory sections. The oxidiser and fuel (propellant) are stored separately in the tanks. Pumps are used to convey the fuel and oxidiser from the tank to the combustion chamber at the required rate through the non-return shutoff valves. The fuel and oxidants are combusted in the combustion chamber, and exhaust gases are expelled out from the nozzle to produce the required pr.opulsive force or thrust. The thrust force is given by Eq. (5.1).

$$F = \dot{m} C_j \tag{5.1}$$

where F is the thrust force, C_j is the exhaust jet velocity and $\dot{m}$ is the mass flow rate of oxidizer and propellant mixture.

The amount of thrust generated is directly proportional to the charge available (or the oxidiser and propellant mixture available) and the exhaust jet velocity. In Eq. (5.1), it is assumed that the combusted gas is expanded in the nozzle completely. The duration of this thrust force is directly proportional to the amount of charge available and the rate of charge consumption.

5.4 THRUST EQUATION

The rocket momentum is given by

$$P = mv_0$$

The forward momentum for a small time is expressed as Eq. (5.2).

$$P + \partial P = (m - \partial m)(v_0 + \partial v) + \partial m(v_0 - v_e)$$

$$P + \partial P = mv_0 + m\partial v - v_e \partial m$$

$$\partial P = m\partial v - v_e \partial m \tag{5.2}$$

Now, thrust force = Momentum/Time

$$F = \frac{dP}{dt} = m\frac{dv_0}{dt} - v_e \frac{dm}{dt}$$

Neglecting gravity, $F = A_e(P_e - P_0)$

$$m\frac{dv_0}{dt} - v_e\frac{dm}{dt} = A_e(P_e - P_0)$$

$$m\frac{dv_0}{dt} = v_e\frac{dm}{dt} + A_e(P_e - P_0)$$

$$F = \dot{m}v_e + A_e(P_e - P_0) \tag{5.3}$$

where F is thrust force, P_e is exit pressure, P_0 is ambient pressure, $\dot{m}$ is mass of charge, v_e is velocity, v_0 is the vehicle velocity and A_e is exit cross-sectional area. Equation (5.3) gives the thrust force generated by the rocket engine.

Specific impulse

Specific impulse, I_{sp}, can be defined as the ratio of thrust force generated by the engine to the unit amount of fuel consumed by the engine. It is used to find the impulsive power of a rocket based on the fuel energy intensity. As discussed in Chapter 1, the specific impulse of a rocket is very less as compared to all other types of propulsion systems. The specific impulse of a rocket is constant for varying or increasing Mach number and compared to other propulsion systems, the specific impulse of a rocket engine is very small. Specific impulse is given by Eq. (5.4).

$$I_{sp} = \frac{F_N}{\dot{m}g} = \frac{F_N}{\dot{w}} \tag{5.4}$$

$$I_{sp} = \frac{1}{\text{SFC}}$$

where I_{SP} is the specific impulse, F_N is the net force, $\dot{w}$ is the rate of propellant weight.

The reciprocal of specific impulse is called *specific fuel consumption* (*SFC*) or *specific propellant consumption*. The higher the specific impulse, the lesser is the requirement of propellant to achieve the required amount of momentum.

Rocket efficiency

Rocket efficiency, η_R is the ratio of total mechanical power output imparted as thrust power to propel the vehicle as opposed to how much it is wasted. It is the ratio of the thrust power generated by the engine to the jet power produced by the exhaust, which is given as follows:

$$\eta_P \text{ or } \eta_R = \frac{\text{Thrust power}}{\text{Jet power}} = \frac{W_F}{W_{\text{jet}}}$$

$$\eta_R = \frac{\dot{m} v_e v_0}{\frac{1}{2}\dot{m}(v_e^2 + v_0^2)} = \frac{2}{\left(\frac{v_e}{v_0} + \frac{v_0}{v_e}\right)}$$

$$\eta_R = \frac{2}{\frac{v_e}{v_0} + \frac{v_0}{v_e}} \tag{5.5}$$

$$\eta_R = \frac{2\alpha}{1 + \alpha^2} \tag{5.6}$$

where v_e is the exhaust velocity, v_0 is the vehicle velocity, α is the velocity ratio. Using Eq. (5.5) and (5.6), the efficiency of the rocket engine can be calculated.

5.5 SOLID PROPELLANT ROCKET ENGINE

As the name indicates, solid propellant rocket engine or motor uses propellant which is in the solid state. The solid rocket motor consists of a combustion chamber and an expansion nozzle. The combustion chamber consists of only propellant and an ignition source, as shown in Figure 5.1.

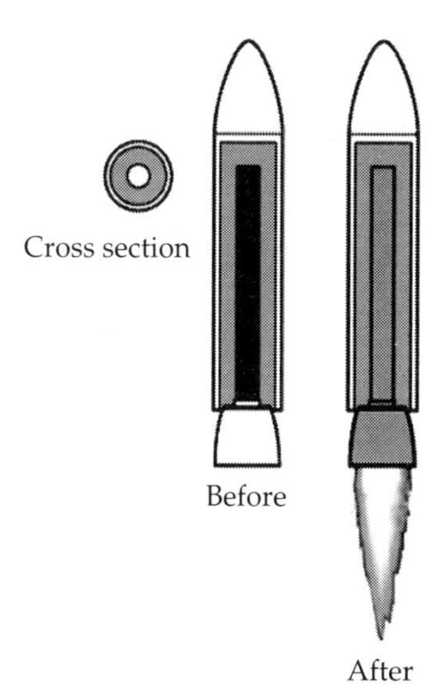

Figure 5.1 Solid rocket motor.

In this type of rocket engine, the entire propellant, which is having its oxidants within the propellant itself, is contained in the combustion chamber. The combustion duration is relatively lower because of higher combustion intensity and higher combustion rate. In this type of rocket engine, the propellant shape is called *grain*.

The igniter is triggered electrically and the propellant burns in steady state; thus, the name 'cigarette burning' is given to this combustion process. The combustion produces gases at very high temperature and pressure. The burning heat is conducted back to the propellant; thus, it is self-sustaining. This process is also called *deflagration*. The combustion products are expanded through the nozzle and the exhaust jet velocity is increased.

The characteristics of solid propellant are as follows:

1. The area of propellant grain determines the magnitude of exhaust gas generation.
2. Thickness or diameter of grain determines the burning rate or time.
3. Propellant grain also acts as insulation for the rocket structure or frame.
4. This cannot be shut down or throttled once combustion starts.

The solid propellant can be aligned inside the combustor in different shape or grain. The grain shape or solid propellant shape inside the combustor depends on the combustion rate and the combustion intensity required. The grain shapes are classified as tubular type, rod and tube type, double anchor type, star type, multi-fin type and double composition type, and so on.

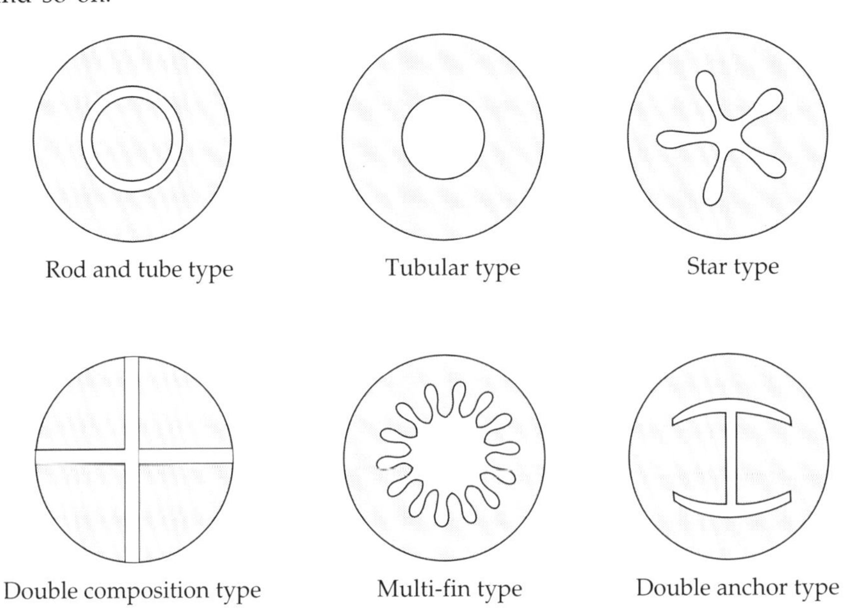

Figure 5.2 Solid propellant grain structure.

In all types of grains, as shown in Figure 5.2, the combustion begins from the core hollow area (except rod and tube type) and the combustion area increases gradually, maintaining the pressure constant. Only in the initial condition of the combustion process, the combustion pressure is less, with less combustion rate in the star, double anchor and multi-fin types of grain. The combustion rate and combustion pressure keep on increasing until there is no change in shape (constant combustion area).

Solid propellant and oxidants

1. Nitrocellulose and nitroglycerine
2. Asphalt and perchlorate
3. Charcoal, sulphur and potassium nitrate
4. Zinc and sulphur
5. Potassium nitrate and sorbitol or sucrose
6. Ammonium nitrate composite (magnesium and aluminum) or ammonium perchlorate
7. CL-20 nitroamine
8. Plastisol

5.6 LIQUID PROPELLANT ROCKET ENGINE

In liquid propellant rocket engine, the propellant is in liquid state. This engine consists of an oxidiser tank, fuel tank, combustion chamber, nozzle, various control valves and motors, as shown in Figure 5.3. The rocket motor or combustion chamber consists of an injection system, which directs the liquid propellant and oxidants to the combustion chamber.

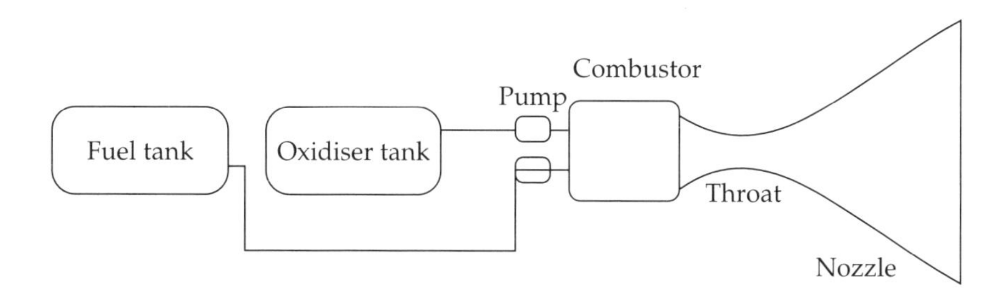

Figure 5.3 Liquid propellant rocket engine.

The combustor and nozzle wall cooling is very necessary to prevent them from the thermal effects where the temperature exceeds 2700°C. The trailing part or nozzle produces supersonic exit velocity in the range of 1500–4500 m/s. Nitrogen is basically used to pressurize the propellant and oxidizer tanks.

The propellant used must be dense, less corrosive, toxic-free, chemically stable and must have high calorific value and smooth ignition. The liquid propellants are classified as monopropellant and bipropellant. *Monopropellants* are those which do not require an oxidiser separately for combustion, for example, hydrazine, nitromethane, and hydrogen peroxide. *Bipropellants* are the liquid fuels which require oxidiser separately for combustion, for example, hydrogen and liquid hydrocarbons.

5.7 HYBRID ROCKET ENGINE

A *hybrid rocket engine* is a combination of both solid propellant and liquid propellant engine to overcome the drawback of each individual engine. In this system, the propellant is in solid state and oxidiser is in liquid state; hence, it is called *hybrid engine* or *hybrid rocket engine*. The solid propellant which is used in this engine has no oxidiser with it like a solid propellant engine. The hybrid engine can be operated intermittently and throttled to a degree of varying the oxidiser flow, as shown in Figure 5.4.

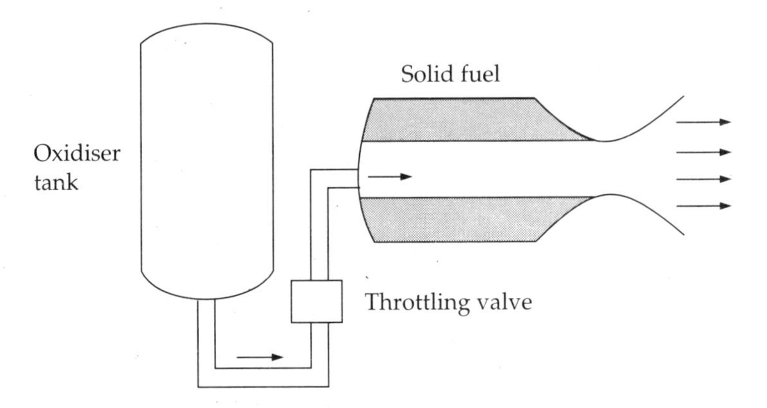

Figure 5.4 Hybrid rocket engine.

It tends to have simplicity and reliability just like the solid propellant engine. Hybrid engines are still under development. The advantage seems to offer good potential for future application.

5.8 ROCKET NOZZLE

Rocket nozzle is a part of thrust chamber which is used to convert the heat energy generated by the propellant combustion into kinetic energy. Most of the rockets use a convergent-divergent nozzle, which is capable of expanding and accelerating the exhaust gas at supersonic speed.

Functions of rocket nozzle

1. To expand the product gases with less total pressure loss
2. To guide the flow with fewer eddies
3. To preheat the fuel before combustion

Nozzle losses

1. Frictional loss: It is due to the viscous force by the hot gases, which is less than 1% of the total loss.

2. Heat transfer loss: Nozzle cools down the hot gases while cooling the walls of nozzle using liquid propellant.

3. Aerodynamic loss: Nozzle exit area is maintained as equal as body area to avoid base drag.

4. Divergence loss: If exhaust gas flow direction is not parallel to the nozzle direction, then such type of loss occurs.

Based on the grain shape or fuel oxidant configuration, different types of convergent divergent nozzles are used in solid and liquid propellant rockets.

An easy-to-manufacture conical nozzle having a simple design (Figure 5.5), is used in most of the small capacity engines. The nozzle has a divergence angle of 10°–20° after throat section. This divergence angle helps in producing a higher amount of thrust with high specific impulse. This leads to the lengthier nozzle to expand the gases, which leads to an increase in weight.

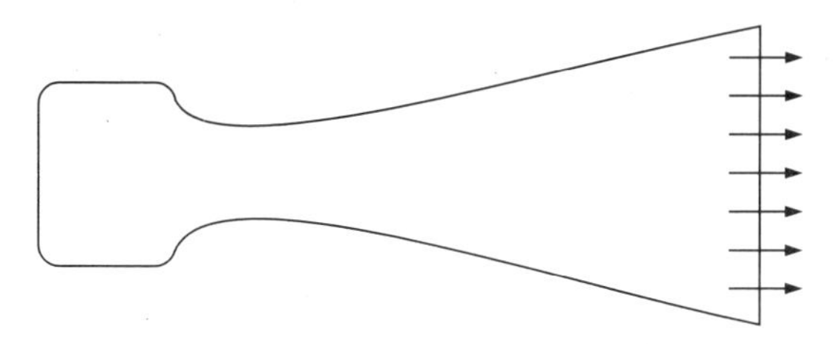

Figure 5.5 Conical type nozzle.

Nowadays, all large capacity rocket engines use contoured nozzle (Figure 5.6). The nozzle diverging angle after throat section is 30°–60°. At the end of the nozzle section, it converges to 2°–10° angle. These nozzles are designed to avoid oblique shock wave. They can achieve their maximum performance level at only a specified design altitude.

Aerospike (Figure 5.7) is the most complex design nozzle. The hot pressurised gas flows around the central body of the nozzle (spike). Spike actually blocks the gas from flowing from the centre of the nozzle and it creates convergent-divergent shape at its periphery with respect to the outer

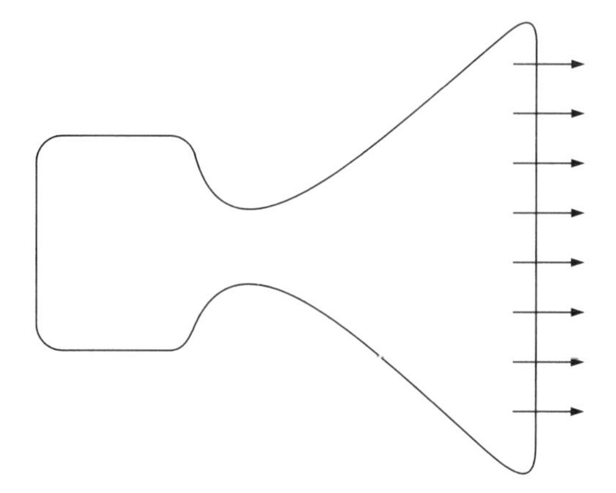

Figure 5.6 Contoured type nozzle.

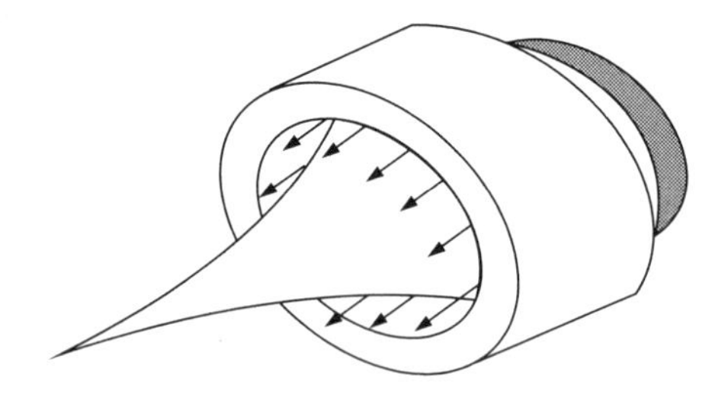

Figure 5.7 Aero-spike type nozzle.

casing of the nozzle. Assuming isentropic expansion, we can say no energy is lost due to the turbulent flow and the nozzle efficiency is maintained.

5.9 ROCKET STAGING

Rockets usually have multiple stages of thrust production systems based on the requirements. Often the first stage is called *booster stage*. It is the biggest stage among all other stages which has greater thrust and specific impulse. In this stage, the combustion process cannot be throttled and not required also, as it is used to propel beyond atmosphere without deceleration; hence, solid propellant is used. For the subsequent stages, comparatively less thrust is required as the rocket goes beyond the atmosphere region (no drag).

Types of rocket staging

There are three types of rocket staging—Tandem, Parallel and Piggyback.

Tandem stage means the rocket engine stages are arranged one above another in a series and only one stage can be used at a time. Once lower stage engine is used completely or if it is not required after some use, it can be detached from rocket and the subsequent next stage will be used to generate thrust. It is also called *series stage rocket,* as shown in Figure 5.8.

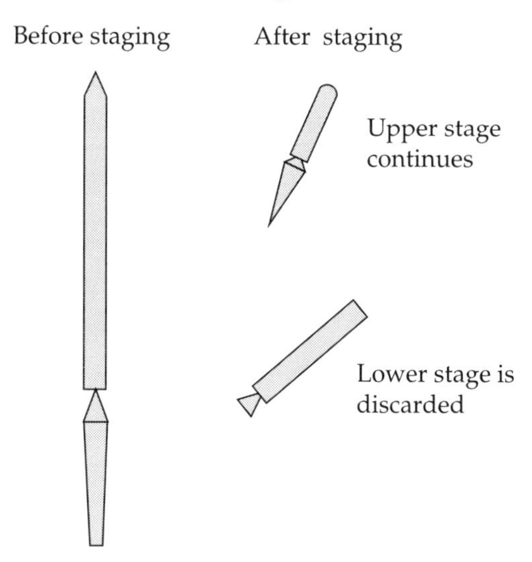

Figure 5.8 Tandem rocket stage.

Parallel rocket stage means the main rocket engine has multiple sub-engines called *stages* or *boosters* around the main engine, as shown in Figure 5.9. The sub-engines (secondary engines) or booster are the solid propellant rocket engines. The sub-engines and main engine together start to generate a huge amount of thrust. Once the rocket booster or sub-engine is used, it is detached from the main rocket engine.

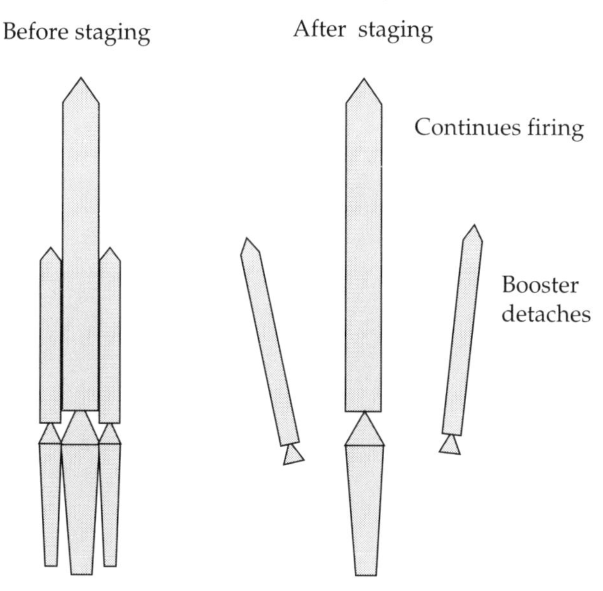

Figure 5.9 Parallel rocket stage.

Piggyback rocket stage is another and more interesting type where the space shuttle is mounted on the main rocket like a piggyback, as shown in Figure 5.10. The main rocket engine is used to carry the space shuttle to a certain altitude, and then, the space shuttle detaches from the rocket and proceeds further using its own engine or thrust generator.

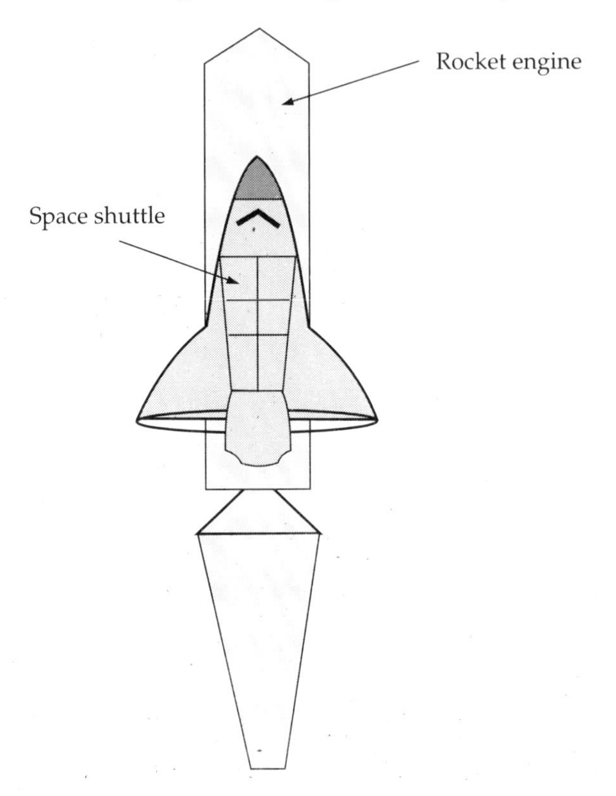

Figure 5.10 Piggyback rocket stage.

IMPORTANT QUESTIONS

1. Define rocket. Explain the working procedure of rocket.
2. Define thrust of rocket engine. How is thrust equation for rocket different from gas turbine engine?
3. Define the specific impulse and efficiency of rocket engine. Compare these parameters with gas turbine engine.
4. Write the characteristics of solid propellant.
5. Explain the three different types of chemical rocket engines.

6. Explain the different grain structures used in solid rocket engine.
7. With a neat diagram, explain the solid rocket booster. Give two examples of solid propellants.
8. With a neat sketch, explain liquid rocket engine. Write the advantages of liquid rocket engine over solid rocket booster.
9. Explain the types of rocket nozzles.
10. Write the functions of rocket nozzle. Explain the nozzle losses.
11. Explain the major types of rocket staging used.

CHAPTER 6

RAMJET AND SCRAMJET ENGINE

OVERVIEW

The chapter throws light upon the meaning of ramjet and scramjet engines and how the compression is done without any rotating component in these engines. It explains about the working principles and operating conditions of ramjet and scramjet engines. The chapter also discusses integral ram-rocket engine.

Thrust force can be produced without having any rotational component in the engine, utilising the atmospheric air at higher Mach values. Combustion process can be carried out when the oxidiser flow velocity is supersonic which does not happen in gas turbine engines. All these statements describe the characters of ramjet and scramjet engine, which is discussed in detail here.

Ramjet engine is an air-breathing engine with no rotating components (turbomachines). This type of engine cannot be used at static condition, as there is no rotating component to pressurise the air for combustion. The Mach values and specific impulse values for the engine have already been described in Chapter 1.

A *scramjet engine* is an advanced version of ramjet engine which has no rotating parts like ramjet engine.

A ramjet engine does not have any compressor or turbine or turbomachine. It relies on its forward motion to push the air through the engine, which means the engine needs to be moving to get the air inside it. Therefore, ramjet cannot produce thrust force at static condition. Ramjet engine is very efficient at Mach 2.5 to 5 and it is inefficient at lower than Mach 2.5 and subsonic speeds. A scramjet engine is efficient at more than Mach 5 speed and inefficient at a lesser value than Mach 5.

The major components of ramjet or scramjet engines are intake, fuel injector, flame stabiliser, combustion chamber and nozzle. As both the types of engines are designed to work at supersonic speed, the intake shape is convergent, which decreases the supersonic flow velocity and increases the pressure. In Ramjet engine, the shape of combustion chamber is convergent and sometimes it is just a parallel duct where the combustion takes place at constant pressure. As the flow comes continuously from the inlet section to the combustion chamber, there should be some obstacle in its path to hold the combustion flame from extinguishing. Such an obstacle is called *flame holder* or *flame stabilizer,* as shown in Figure 6.1. The nozzle used in this type of propulsion system is convergent-divergent type nozzle as the nozzle pressure ratio is more than 5. A ramjet can be integrated with rocket engine as solid fuel ramjet engine or liquid fuel ramjet engine.

6.1 RAMJET ENGINE

Ramjet propulsion system (Figure 6.1) can be used efficiently at supersonic speed. The engine uses external thrust force (another propulsion system) to gain the initial momentum (supersonic speed) so that the atmospheric air can enter the intake section at supersonic velocity. The air is compressed at the intake section or at the inlet as internal compression and external compression because of inlet shape. Internal compression is nothing, but the air compression inside the engine or intake, and external compression means compression of air outside the intake section.

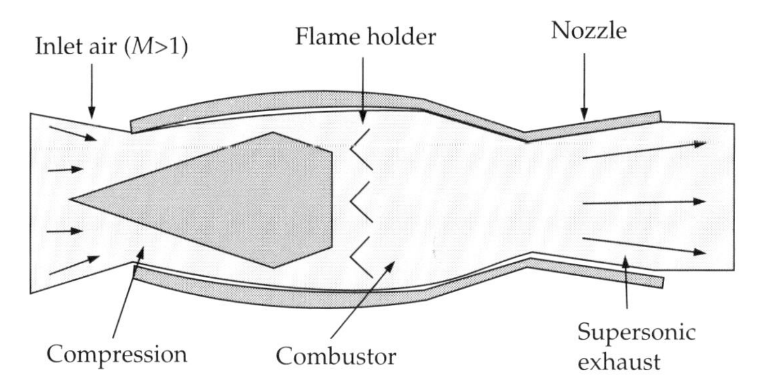

Figure 6.1 Ramjet engine.

The supersonic velocity of air creates shock wave at the inlet section and the shock wave, which is near the lip of the inlet, gets inside it and undergoes multiple reflections. The air flowing through the shock wave has changes in its properties like increase in pressure, decrease in velocity and also increase in temperature and density. This phenomenon is called *internal compression*. The inlet cross-section makes the flow get compressed before it reaches the inlet section, as the air flow velocity is supersonic. This is called *external compression*. The internal and external compression is true for scramjet also.

The high-pressure air is supplied to the combustion chamber where the velocity is comparatively greater than the required, which leads to incomplete combustion with less combustor cooling and also friction because of the flow inside the combustion chamber. All these lead to pressure loss inside the combustion chamber. Compression is initially achieved by shock wave followed by subsonic diffusion. The uninstalled thrust equation for a ramjet engine is same as that for a gas turbine engine, i.e., Eq. (6.1).

$$F = ma = (\dot{m}_e v - \dot{m}_0 u) + A_e (P_e - P_0) \tag{6.1}$$

where, F is the thrust, m is the mass of fuel, a is the acceleration, $\dot{m}_e$ is the exit mass flow rate, v is the exit jet velocity, $\dot{m}_0$ is the inlet mass flow rate, u is the vehicle velocity, A_e is the exit area of nozzle, P_e is exit pressure of jet and P_0 is the ambient pressure.

For these conditions, the convergent-divergent nozzles are designed so that it will be choked at the throat which is a critical condition of the nozzle. For optimal case and fully expanded nozzle, we have thrust [Eq. (6.2)] as follows:

$$F_N = \dot{m}_e v_e - \dot{m}_0 v_0 = (\dot{m}_{\text{air}} + \dot{m}_{\text{fuel}})\, v_e - \dot{m}_{\text{air}} v_0$$

$$F_N = \dot{m}_{\text{fuel}}[(\text{AFR} + 1)\, v_e - (\text{AFR})\, v_0] \tag{6.2}$$

where F_N is the net thrust, $\dot{m}_{\text{fuel}}$ is the mass flow rate of fuel, AFR is the air-fuel ratio, v_e is the exit velocity and v_0 is the vehicle velocity.

Specific impulse is given by Eq. (6.3).

$$I_{sp} = \frac{F_N}{\dot{m}_{\text{fuel}}\, g} = \frac{[(\text{AFR} + 1)\, v_e - (\text{AFR})\, v_0]}{g} \tag{6.3}$$

In ramjet engine fuel is injected in a similar manner as done in the gas turbine engine, but in scramjet engine, the fuel is injected from the walls or using some strut arrangements which have a flame holder structure. Generally, hydrogen is used as fuel, but kerosene-based hydrocarbon fuel can also be used such as JP-10, JP-4 and JP-5. Better volume utilisation can be achieved by using high-density liquid fuel. Some part of air is often directed radially into the combustion chamber of ramjet engine so that the flame stabilisation can be enhanced by creating toroidal vortex recirculation.

6.2 SCRAMJET ENGINE

A *scramjet engine* is a supersonic combustion ramjet engine, which we can say as improved ramjet engine also. In a scramjet engine, the air flows through the intake or inlet of engine which result increase in pressure and temperature and decrease in flow velocity. Combustion of the air-fuel mixture occurs at supersonic speed. This makes the engine more efficient by having a high value of momentum of air throughout the engine without slowing down the air as in ramjet engine. In scramjet engine (Figure 6.2), the requirement of the combustion process is comparatively lower because of the type of fuel used (hydrogen).

The initial momentum of the engine is provided by some external propulsion source to reach the required velocity. The atmospheric air enters the intake section and gets compressed by the internal compression and the external compression. This compression can also be called *ramming*. The flow after intake section remains supersonic, with more pressure and temperature values compared to the intake entry condition.

The initial flow conditions from the intake section to the combustor are sufficient to have complete combustion of hydrogen-air mixture efficiently without slowing the air speed to subsonic. The combustion chamber is divergent in shape, which prevents the excessive pressure increase and accelerates the combusted gas, as shown in Figure 6.2. The combustion can be made more effective by adding oxygen as the oxidiser. The addition of oxygen is 1.5 times more effective than adding hydrogen in this engine. The performance of a combustor can be improved by either a hot combustor inlet temperature or by incorporating an advanced ignition system into the scramjet that effectively eliminates the ignition delay. The combusted high-velocity gases are expelled out from the nozzle at much higher velocity. Because of all these individual processes, the scramjet engine can travel at a higher value of supersonic speed and beyond it as well.

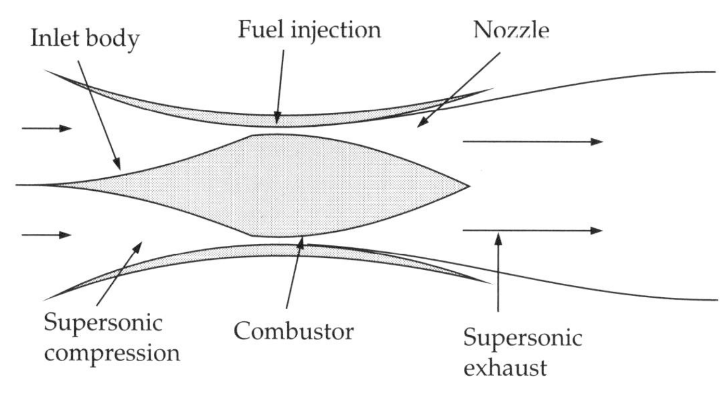

Figure 6.2 Scramjet engine.

The combustion depends on the enthalpy, which is proportional to pressure, air-fuel ratioand inlet temperature of fuel and air. An aircraft powered by scramjet engine is capable of achieving Mach 10 and can reduce the transient flight time. The area under this path is directly proportional to the energy added and can be terminated at any point along the curve. Heat can be added until the maximum entropy condition occurs at combustion zone. At this point, the flow is decelerated to its maximum limit and it is thermally choked. This imposes maximum heating limitation in scramjets. An increase in entropy occurs at intake due to total pressure losses associated with this diffuser.

6.3 INTEGRAL RAM-ROCKET ENGINE

The rocket which is designed to use atmospheric air as oxidant and compressed by internal and external ramming is called *integral ram-rocket engine*. This can be solid fuel or liquid fuel ram-rocket engine. This system can be used when the rocket is within the atmosphere with initial propulsion by some propulsion systems.

In *solid fuel ram-rocket engine,* the air is sucked from the atmosphere and it is rammed to a certain pressure and directed to the combustion chamber where it comes in contact with the solid fuel and ignition energy. A continuous supply of air helps in maintaining continuous combustion inside the chamber. This avoids the usage of bipropellant. The free air helps in achieving complete combustion of the fuel.

In *liquid ram-rocket engine,* the liquid fuel is stored in a separate tank and the oxidiser required for combustion process is taken from the atmosphere. The atmosphere pressure is not enough to carry the combustion; hence, it is rammed before taking it inside the combustion chamber. The liquid fuel is sprayed using fuel injector. The air is directed in such a way that it mixes with fuel and further helps in maintaining the combustion by creating the turbulent effect.

The ramjet engine integrated with the gas turbine engine also exists and is called *turbo ramjet engine*. In this type of engine, the inlet part is like ramjet engine followed by the turbomachines and combustor of gas turbine engine. The ramjet inlet helps in compressing the air, which reduces load from the compressor. Because of this integrity, this engine can fly at very high velocity (Mach 3) efficiently.

IMPORTANT QUESTIONS

1. Define ramjet and scramjet engines. Do these engines follow Brayton cycle or Humphrey cycle?
2. With a neat diagram, explain the working of ramjet engine.
3. Write the differences between ramjet engine and scramjet engine.
4. With a neat diagram, explain scramjet engine.
5. With respect to cross-sectional area, explain the differences between ramjet and scramjet engine.
6. Can a ramjet engine be integrated with rocket engine and also with gas turbine engine? Justify.

CHAPTER 7

PULSE DETONATION ENGINE

OVERVIEW

The chapter begins with the meaning of pulse detonation engine and its thermodynamic cycle. It explains about the major components of pulse detonation engine, working principle of the engine, classification and some engine aspects like detonation, deflagration, cell size, shock, flame interaction, and so on. This chapter is useful for understanding the combustion in constant volume heat addition process with a shock wave.

Pulse detonation engine is an unsteady propulsive device with highest specific impulse and specific thrust among all other steady and unsteady chemical propulsion systems. It works on Humphrey cycle (constant volume heat addition process). This propulsion technology involves detonation after combustion of the air-fuel mixture to produce thrust more efficiently. This engine has got the capacity to revolutionise the aviation domain.

Pulse detonation engines are an extended version of pulse jet engines and they share many similarities. However, there is one important difference between them—pulse detonation engine detonates rather than deflagrates. The operation of pulse detonation engine can be explained very well through Chapman–Jouguet condition, i.e., air-fuel mixture injected into a hollow tube will undergo combustion in the presence of ignition source and get transitioned from deflagration to detonation (strong and weak points of deflagration and detonation on *P-V* chart), if the flow is choked partially in the hollow tube and combusted products are expelled out from the tube at supersonic speed.

The major components of pulse detonation engine include detonation tube, which is closed at one end and open at another end. Towards its closed side, mixture inlet is provided to have charge inside the engine. From the open end, the combusted gas is taken out. The injector injects the pre-mixture of air and fuel into the detonation tube from the inlet provided at the closed end. *Shchelkin spiral* is a dead component inside the detonation tube, which aids in deflagration to detonation transition. Ignition source provides an electric spark of a certain length for combustion. In a few pulse detonation engines, Pre-heaters are also used, which preheat the oxidiser or propellant or both based on the requirement and design conditions, as shown in Figure 7.1.

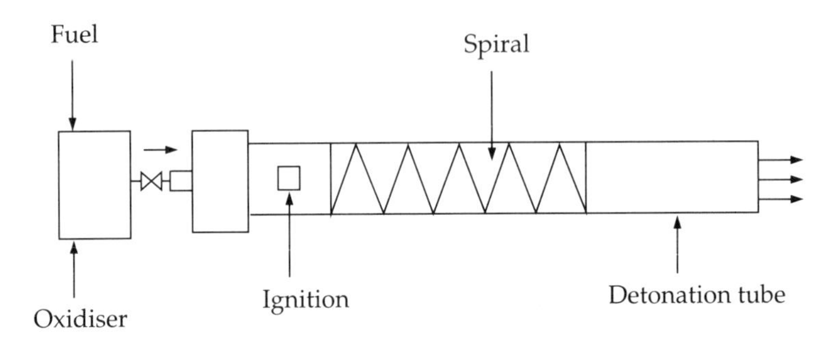

Figure 7.1 Pulse detonation engine.

7.1 CLASSIFICATION OF PULSE DETONATION ENGINE

1. Based on the type of fuel used
   (a) Gaseous fuel pulse detonation engine
   (b) Liquid fuel pulse detonation engine
   (c) Hybrid fuel pulse detonation engine
2. Based on blocking object
   (a) Shchelkin spiral
   (b) Orifice or perforated plate
   (c) Combination of orifice or perforated plate with Shchelkin spiral
3. Based on preheat setup
   (a) Preheated oxidiser
   (b) Preheated propellant
4. Based on charge supply
   (a) Valve type (rotary valve)
   (b) Valveless type

Pulse detonation engine can also be categorised mainly into three broad categories, namely, pure, combined cycle and hybrid pulse detonation engine. *Pure pulse detonation engine* is one that relies only on a pulse detonation engine which consists of detonation tube, an inlet and a nozzle. A *combined cycle pulse detonation engine* uses different speed ranges. This includes the combination of pulse detonation engine with ramjet or scramjet. *Hybrid pulse detonation engine* is a combination of pulse detonation engine with turbofan or turbojet engine.

In pulse detonation engine, the oxidiser and propellant both can be gaseous, for example, H_2 and O_2. For such a combination, the deflagration to detonation transition length is very small; hence, the efficiency is more. If liquid oxidiser and propellant are used in pulse detonation engine, then the storage, throttling and frequency matching with the detonation and ignition system will be difficult. To overcome these drawbacks, hybrid fuel pulse detonation can be used where the propellant can be liquid, i.e., hydrocarbon and oxidiser can be gaseous, i.e., pressurised air or oxygen.

In the hybrid fuel pulse detonation engine, the storage, throttling and achieving detonation in small length is adaptable. In liquid and hybrid fuel pulse detonation engine, there is one more problem in combustion, i.e., two-phase problem. The two-phase problem affects detonation performance, ignition performance, delay and rapid variation in combustion temperature.

As mentioned in the definition, pulse detonation engine is an unsteady detonation engine. This detonation can be achieved by varying the inside cross section of the combustion section. To vary the cross-sectional area, Shchelkin spiral or orifice plate or perforated plate (mesh plate) is used. For gaseous fuel pulse detonation engine, perforated or orifice plate is enough to achieve the detonation, whereas in liquid and hybrid type, combination of Shchelkin spiral and an orifice or perforated plate is used. The Shchelkin spiral or orifice or perforated plate creates blockage inside the detonating tube, thus increasing the flow velocity by making it flow through the convergent duct. This enhances the efficiency and output.

As discussed earlier, the two-phase problem in liquid type and hybrid type pulse detonation engine can be resolved using a preheating method. Preheating the fuel or air using a heat exchanger helps in avoiding the two-phase problem. If preheating of fuel is planned, then care has to be taken to avoid the auto-ignition problem. The auto-ignition problem can be rectified by using Nitrogen to compress the fuel in the tank, rather than air. If preheating of oxidiser is planned, then the two-phase problem will be very less in running the liquid or hybrid pulse detonation engine. The performance of the engine is better if oxidiser is preheated than the propellant preheating.

In a few types of pulse detonation engine, the air-fuel mixture is throttled using some electromechanical valves (for example, solenoid valve) with a continuous ignition source and in some other type of pulse detonation engine, the air-fuel mixture is supplied continuously with a discrete ignition source. If the air supply to the pulse detonation engine is continuous, then to make it discrete rotary valves can be used, and it is generally called *valve type pulse detonation engine*.

Some of the commonly used fuels to study the pulse detonation engine are LPG, hydrogen, propane, gasoline, aviation kerosene, etc. The performance level of pulse detonation engine is high when oxidiser is oxygen as compared to air. For oxygen, the transition from deflagration to detonation can be with or without blocking object with less transition distance. The current combustion and system models predict very high propulsion efficiencies for pulse detonation engine and good thrust characteristics from low subsonic to high supersonic flight regimes.

7.2 CHAPMAN–JOUGUET (C–J) CONDITION

The C–J condition states that the detonation propagates at a velocity at which the reacting gases just reach sonic velocity. Chapman stated that there exists a minimum velocity at which detonation can occur and it is thermodynamically tied to the properties of the burning gas. Jouguet established the relation that the detonation wave velocity is equal to sound velocity or the combusted gas in which it propagates. It holds approximately in detonation wave of high explosives.

In Figure 7.2, the deflagrations to detonation transition points are shown in two different locations. When oxygen is used as oxidant, then strong detonation region is achieved. The strong detonation point, which has got a high amount of pressure, is converted into velocity shown in strong deflagration region, and weak detonation point, which has got less pressure, is converted into low-velocity shown in weak deflagration region.

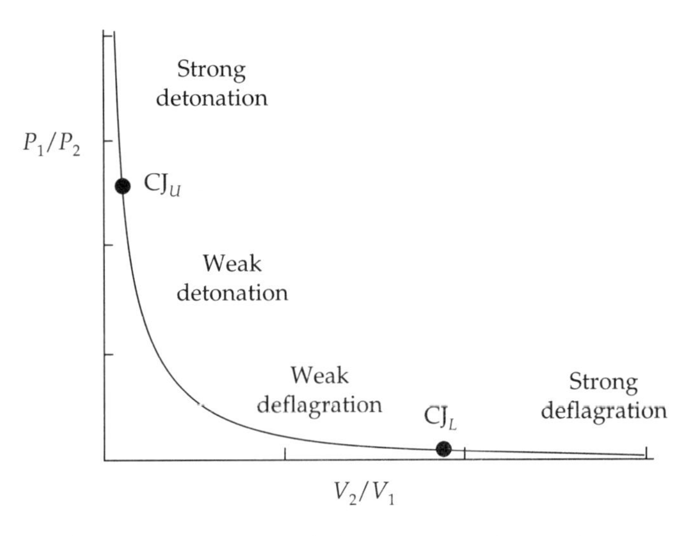

Figure 7.2 Chapman–Jouguet condition for pulse detonation engine.

The shock in the pulse detonation engine rises the pressure to ZND points, commonly called *von Neumann spike*. In the C–J model, the detonation wave or shock wave is closely coupled to a thin flame or combustion wave, front or combustion region. If the shock wave is followed by a combustion wave, then it is known as *stable detonation*, and if the shock wave follows the combustion wave, then it is called *unstable detonation*. The conservation condition requires that the final state lies on both the reactive Hugoniot curve. The tangent points are called *the upper and lower C–J points*.

7.3 COMBUSTION WAVE THEORY

For a liquid fuel pulse detonation engine, stable detonation occurs only at the upper C–J point. The gaseous wave speed of the upper C–J point is the primary metric used in pulse detonation engine research to confirm the existence of detonation wave. The upper C–J wave speed is used to determine deflagration to detonation transition time, detonation distance and the percentage of ignition resulting in detonations during this effort.

7.4 DEFLAGRATION TO DETONATION TRANSITION

When the air-fuel mixture is ignited, the deflagration or subsonic combustion of the mixture takes place. When these combusted gases are blocked partially using Shchelkin spiral and orifice plate or perforated plate, the deflagration to detonation transition takes place, which is at supersonic speed. The detonation energy release rate is much higher than in deflagration. Engines utilising detonation have higher thermodynamic efficiency and they are easier to scaling as compared to conventional engines which use deflagration combustion. Deflagration is a surface phenomenon where the hot products heat up the next layer of reactants along the surface for combustion at a subsonic velocity. A detonation is a rapid violent chemical reaction that proceeds through the products towards the reactants at supersonic velocity.

The thermodynamic cycle of pulse detonation engine is Humphrey cycle, as mentioned earlier. It is considered as modified Brayton cycle where the constant pressure heat addition is replaced by constant volume heat addition and the combustion exit temperature is higher in Humphrey cycle than the Brayton cycle. The working cycle, as shown in Figure 7.3, is as follows:

1. Reversible adiabatic (isentropic) compression of the incoming gas: During this process, the stagnation pressure and temperature increase because of the work done on the charge. The charge is compressed in its tank using external high-pressure air or nitrogen. As the charge gets pressurised, the pressure increases maintaining constant entropy.

2. Constant volume heat addition: In this process, heat is added while the gas is at constant volume. In most cases, Humphrey cycle engines consider open cycle, which means the specific volume remains constant throughout the heat addition process. In pulse detonation engine, from the ignition time to detonation, the transition time is referred to as *heat addition process.*

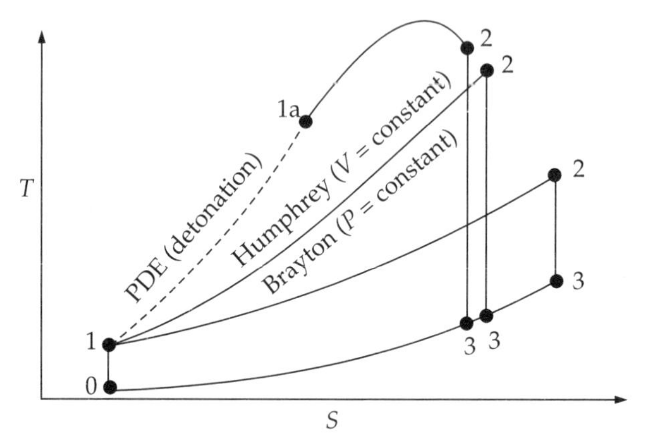

Figure 7.3 Brayton and Humphrey cycle.

3. Isentropic expansion of the gas: In this process, the stagnation temperature and pressure decrease because of the work extracted from the gas by the flow expanding the device. In this process, the pressure decreases maintaining constant entropy. In pulse detonation engine, the detonation shock with combusted gas starts moving to the open end of tube, which is nothing but the expansion of the combusted gas.

4. Constant pressure heat rejection: In this process, heat is removed from the working fluid while the fluid remains at constant pressure. In open cycle engines, this process usually represents expulsion of the gas from the engine, where it quickly equalises to ambient pressure and slowly loses heat to the atmosphere, which is considered as an infinitely large reservoir for heat storage with constant pressure and temperature.

One-dimensional detonation wave is described well using the ZND model, but an actual detonation wave is multi-dimensional in behaviour. Within the narrow channel, detonation waves have primarily two-dimensional phenomenon. A fully developed detonation wave traversing through the reacting mixture produces repeating structures, known as *cells.* The path traversed by the triple point binds the cell structure.

The *triple point* is the location where the Mach stem, incident shock and reflected shock intersect. The *cell size* (λ) is defined as the height of the cell structure and is related to the direct detonation initiation energy, as shown in Figure 7.4. The *direct detonation initiation energy* is the experimentally determined energy required by a combustion system to initiate the detonation directly.

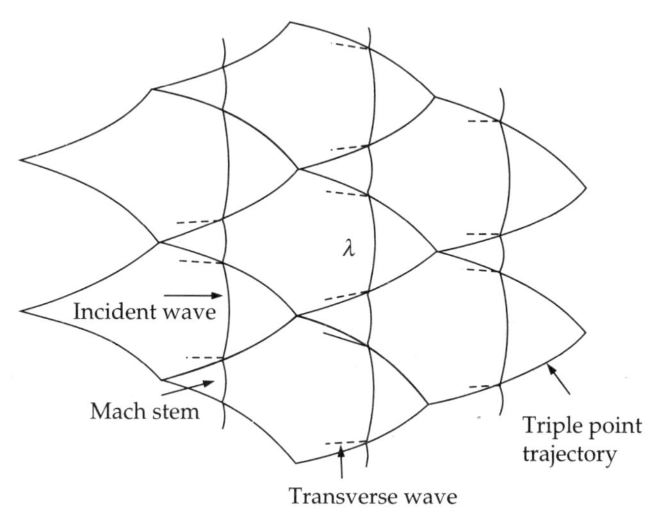

Figure 7.4 Cell size of the propellant.

In pulse detonation engine, the overall weight of the engine is less, fuel consumption is less and the detonation shock is used to generate thrust. All these together make this engine more efficient. The major disadvantages of pulse detonation engine are vibration and noise.

IMPORTANT QUESTIONS

1. Define pulse detonation engine. Explain how it is different from pulse jet engine.
2. Explain the thermodynamic cycle of pulse detonation engine.
3. Explain the pulse detonation engine working procedure with essential components.
4. Explain the classification of pulse detonation engine.
5. Define deflagration and detonation.
6. Explain Chapman-Jouguet condition.
7. Explain how deflagration combustion is transformed into detonation combustion in pulse detonation engine.
8. What do you mean by cell size of propellant?
9. Specific impulse in pulse detonation engine is more than any chemical propulsion system. Why?

SUGGESTED READINGS

Babu, V., *Fundamentals of Propulsion*, Ane Books, New Delhi, 2009.

Cohen, H., Rogers, G.F.C., Saravanamuttoo, H.I.H. and Staznicky, P.V., *Gas Turbine Theory*, 6th ed., Pearson Education, England, 2009.

Heywood, John B., *Internal Combustion Engine Fundamental,* McGraw-Hill, USA, 1988.

Liberman, Michael A., *Introduction to Physics and Chemistry of Combustion,* Springer, Germany, 2008.

Mattingly, J.D., *Elements of Gas Turbine Propulsion,* McGraw-Hill, AIAA, USA, 1996.

INDEX